Practice and Procedure
for the Quantity Surveyor

Also available

Elements of Quantity Surveying
Eighth Edition
Christopher J. Willis and Donald Newman

Specification Writing for Architects and Surveyors
Tenth Edition
Christopher J. Willis and J. Andrew Willis

Cost Planning of Buildings
Sixth Edition
Douglas J. Ferry and Peter S. Brandon

Principles of Building Economics
John Raftery

Quantity Surveying Techniques
New Directions
Edited by Peter S. Brandon

Risk Management and Construction
Roger Flanagan and George Norman

Practice and Procedure for the Quantity Surveyor

Tenth Edition

Christopher J. Willis
FRICS, FCIArb

Allan Ashworth
MSc, ARICS

and

J. Andrew Willis
BSc, ARICS

**Blackwell
Science**

© A.J. Willis & C.J. Willis 1963, 1966,
   1969, 1972, 1975, 1980
© C.J. Willis & A. Ashworth 1987
© C.J. Willis, A. Ashworth & J.A. Willis
   1994

Blackwell Science Ltd
Editorial Offices:
Osney Mead, Oxford OX2 0EL
25 John Street, London WC1N 2BL
23 Ainslie Place, Edinburgh EH3 6AJ
350 Main Street, Malden
   MA 02148 5018, USA
54 University Street, Carlton
   Victoria 3053, Australia
10, rue Casimir Delavigne
   75006 Paris, France

First edition published by Crosby
   Lockwood and Son Ltd 1951
Second edition 1957
Third edition 1963
Fourth edition 1966
Fifth edition (metric) 1969
Sixth edition 1972
Seventh edition by Crosby Lockwood
   Staples 1975
Eighth edition by Granada Publishing
   Limited 1980
Reprinted 1981
Reprinted by Collins Professional and
   Technical Books 1985
Ninth edition 1987
Reprinted by BSP Professional books 1990,
   1992
Tenth edition by Blackwell Science 1994
Reprinted 1996, 1999

Set by DP Photosetting, Aylesbury, Bucks
Printed and bound in Great Britain at the
Alden Press Limited, Oxford
   and Northampton

DISTRIBUTORS
   Marston Book Services Ltd
   PO Box 269
   Abingdon, Oxon OX14 4YN
   (Orders: Tel: 01235 465500
            Fax: 01235 465555)

USA
   Blackwell Science, Inc.
   Commerce Place
   350 Main Street
   Malden, MA 02148 5018
   (Orders: Tel: 800 759 6102
                 781 388 8250
            Fax: 781 388 8255)

Canada
   Login Brothers Book Company
   324 Saulteaux Crescent
   Winnipeg, Manitoba R3J 3T2
   (Orders: Tel: 204 837 2987
            Fax: 204 837 3116)

Australia
   Blackwell Science Pty Ltd
   54 University Street
   Carlton, Victoria 3053
   (Orders: Tel: 03 9347 0300
            Fax: 03 9347 5001)

A catalogue record for this title
is available from the British Library

ISBN 0-632-03831-4

Library of Congress
Cataloging-in-Publication Data

Willis, Christopher James.
   Practice and procedure for the quantity
surveyor/Christopher J. Willis, Alan
Ashworth, and J. Andrew Willis. — 10th ed.
      p. cm.
   Includes bibliographical references and
index.
   ISBN 0-632-03831-4
   1. Building—Estimates—Great Britain.
1. Ashworth, A. (Allan) II. Willis, J. Andrew.
III. Title.
TH435.W6853   1994
6921'.5'0941 — dc20   94-16578
                                      CIP

To the memory of Arthur Willis,
scholar and quantity surveyor

Contents

Preface

This book was first published in 1951, over 40 years ago, and in each succeeding edition the text has endeavoured to reflect the changing scene of practice. The fact that a tenth edition is required shows that there is still a need for such a book. However, the changes in practice over the past few years pose the question: 'Is a textbook that purports to teach good practice as envisaged by a formal contract the right method of presentation when in practice, because of a variety of circumstances, fee competition, market demand and other such inhibitions, different practices are adopted?'

In attempting to answer that question we have taken soundings from practitioners and from the teaching establishments. The almost unanimous view – which our own experience bears out – is that short cuts have to be taken when circumstances so dictate; but woe betide the practitioner who takes the short cut without knowing the long way round. The person that does is unable to weigh up the risk that they are taking, and if the result lays them open to a charge of negligence then they could well be calling on their professional indemnity insurers.

We have therefore decided to present this new edition in a form that we hope guides students and young practitioners into the ways of practice that reflect the role of the quantity surveyor as it is generally accepted. We are very conscious of the reasons why deviation from the accepted path is sometimes appropriate, and in consequence the warnings referred to above are highlighted where possible.

There is no hard and fast dividing line between 'practice' and 'procedure'. Generally speaking, 'procedure' may be regarded as being governed by rules over which we have little or no control, but by which we must learn to do our work correctly. 'Practice' is more the way we do it, representing decisions that are left (more or less) to our own choice. That a quantity surveyor must give notice to the contractor before commencing measurement of variations is a matter of procedure laid down by both the Standard Form of Contract and the Government Conditions of Contract; how it is done – by telephone, letter or facsimile for example – is a matter of practice.

The format of this edition follows that of previous editions. We have reflected the seemingly endless amendments to standard forms of contract and, where appropriate, have made reference to new forms that have been published since the last edition of this book. We have also brought the bibliographies up to date.

One change that we have made is to direct the emphasis away from private practice, recognizing that the majority of graduates now go to commercial organizations rather than to private practice. This has meant enlarging the text on the role of the quantity surveyor in contracting organizations and putting less emphasis on guidance on dealing with law and the quantity surveyor, as this subject has grown to such an extent that study of separate textbooks is needed to do justice to this aspect of a quantity surveyor's work. One new chapter appears on arbitrations and alternative dispute resolution: two areas in which quantity surveyors are becoming more and more involved.

The book commences with an outline of quantity surveying practice covering the structure of the industry and the work of the quantity surveyor in the various fields, and covering the law and computer applications in so far as they affect practice. The second section covers procedure as regards documentation, cost control and office and site applications. The book ends with chapters on specializations that are applicable to quantity surveyors in whatever forum they practice.

The first edition of this book reflected the practical experience of Arthur Willis in a way that was intended to be helpful to the student and the young practitioner. We can only hope that we have maintained the high ideals of a man so highly regarded by the profession of his time and we are honoured to dedicate this book to his memory.

Christopher Willis
Allan Ashworth
Andrew Willis

Acknowledgements

We are grateful to Dr David Chappell for his agreement for use in this book of parts of the texts and flow diagrams contained in *The Architect in Practice*. We are also grateful to the assistance given to us by our colleagues in private and public practice and in construction firms.

We are grateful to the Royal Institute of British Architects, the Royal Institution of Chartered Surveyors and the National Joint Consultative Committee for permission to reproduce their standard forms and specimen letters.

Abbreviations

ACA	Association of Consultant Architects
ACE	Association of Consulting Engineers
ADR	alternative dispute resolution
AER	*All England Law Reports*
APC	assessment of professional competence
BCA	British Cement Association
BCIS	Building Cost Information Service
BEC	Building Employers Confederation
BLR	*Building Law Reports*
BMI	Building Maintenance Information
BPF	British Property Federation
BPIC	Building Project Information Committee
BRE	Building Research Establishment
BSI	British Standards Institution
CAD	computer-aided design
CASLE	Commonwealth Association of Surveying and Land Economy
CAWS	Common Arrangement of Work Sections for Building Works
CCPI	Coordinating Committee for Project Information
CIArb	Chartered Institute of Arbitrators
CIB	International Council for Building Research Studies and Documentation
CIC	Construction Industry Council
CIOB	Chartered Institute of Building
CIRIA	Construction Industry Research and Information Association
CISC	Construction Industry Standing Conference
CITB	Construction Industry Training Board
CLD	*Construction Law Digest*
CPD	continuing professional development
CPI	Coordinated Project Information
EC	European Community
EDC	Economic Development Committee
EDI	electronic data interchange
ENBRI	European Network of Building Research Institutes
EU	European Union
FCEC	Federation of Civil Engineering Contractors
FIDIC	Conditions of Contract (International) for Works of Civil Engineering Construction

GC/Wks	Government Works Contract
GDP	Gross Domestic Product
HBF	House-Builders Federation
HMSO	Her Majesty's Stationery Office
ICE	Institution of Civil Engineers
IChemE	Institution of Chemical Engineers
IFC	Intermediate Form of Contract
IT	information technology
JCT	Joint Contracts Tribunal
NALGO	National Association of Local Government Officers
NBS	National Building Specification
NCG	National Contractors Group
NCVQ	National Council for Vocational Qualifications
NEC	National Economic Council
NEDC	National Economic Development Council
NEDO	National Economic Development Office
NJCBI	National Joint Council for the Building Industry
NJCC	National Joint Consultative Committee
NSC/A	Standard Form of Nominated Sub-Contract – Agreement
NSC/C	Conditions
NSC/N	Nomination
NSC/T	Tender
NSC/W	Warranty
QA	quality assurance
R&D	research and development
RIBA	Royal Institute of British Architects
RICS	The Royal Institution of Chartered Surveyors
SCQSLG	Society of Chartered Quantity Surveyors in Local Government
SMM	Standard Method of Measurement of Building Works
TRADA	Timber Research and Development Association
UCATT	Union of Construction, Allied Trades and Technicians
VAT	value added tax

Chapter 1

Structure of the construction industry

Introduction

The construction industry includes the activities of the separate sectors of activity of building, civil engineering and heavy engineering. However, the demarcation between these different areas of work is blurred. The construction industry is concerned with the planning, regulation, design, manufacture, fabrication, erection and maintenance of buildings and other structures. Its projects can vary from work worth only a few hundred pounds to major schemes costing several million pounds. While the principles of execution of such projects are similar, the scale, complexity and organization will vary enormously.

The UK Economy

Before considering the structure of the construction industry it is worth examining the trends in the economy in which that industry is placed. In the UK during the last decade of the twentieth century the dominant issues are:

- uncertainty in business cycles, resulting in shorter booms and slumps (3–4 years);
- high levels of unemployment;
- low rates of interest;
- falling rates of inflation;
- reduced outputs in manufacturing;
- shift from a manufacturing to a professional services base;
- emphasis upon market economics;
- government disengagement from industry through privatizations;
- reduction in the influence of trade unions;
- changing age structure of the population;
- reduction in the number of young people;
- regional variation changes.

Characteristics of the construction industry

The construction industry is an important industry in any economy. In the UK it accounts for about 6% of GDP, providing for over half the fixed capital investment. This percentage is subject to annual variations, and construction's share of the national output has declined over the past 20 years. The industry is also a major

employer of labour, from the unskilled through to the high-technology professional. The fortunes of the construction industry provide one of the best indicators
of a country's economic performance. A busy construction industry generally
represents a buoyant economy: the pattern of housebuilding, for example, is a
useful predictor of future booms and slumps throughout the nation.

The industry is characterized by the following:

- The physical nature of the product.
- The product is normally manufactured on the client's premises, (that is, the
  construction site).
- Many of its projects are one-off designs in the absence of a prototype model.
- The traditional arrangement separates design from manufacture.
- It produces investment rather than consumer goods.
- It is subject to wider swings of activity than most other industries.
- Its activities are affected by the vagaries of the weather.
- Its processes include a complex mixture of different materials, skills and
  trades.
- Typically, throughout the world, it includes a small number of relatively
  large construction firms and a very large number of small firms.
- The smaller firms tend to concentrate on repair and maintenance.

At its best the UK construction industry achieves standards of performance
equal to any in the world.

Size and complexity

In 1990, the total value of construction output in the UK reached £45bn (Table
1.1). In addition to this, UK consultants and contractors undertook about 10–15%
of their annual turnover overseas, notably in the Far East, Africa and the
European Union (EU). The industry experienced rapid growth in the late 1980s,
but suffered a serious recession in the early 1990s, the impact of which would
have been more serious had it not been for work already in progress, including
the number of major schemes that were then under construction.

Table 1.1   Type of work undertaken by contractors.

Firm's size (employees)	New work		Repairs and maintenance	Totals
	Housing		Non-housing	
	£bn	£bn	£bn	£bn
1–7	3.88	5.63	1.95	11.46
8–14	7.85	2.49	3.12	13.46
15–1199	9.44	0.94	1.75	12.13
1200 and over	6.76	0.23	0.66	7.65
Totals	27.93	9.29	7.48	44.70

(Source: Housing and Construction Statistics)

Table 1.2   The Western European construction market.

Country	Population (millions)	Share of construction (%)
West Germany	61.5	30.3 + East Germany 77.2m
Italy	58.5	16.6
France	55.5	16.4
UK	55.4	10.7
Switzerland	6.4	5.2
Sweden	8.3	4.5
Netherlands	14.3	4.4
Austria	7.6	3.8
Denmark	5.1	3.0
Norway	4.1	2.7
Belgium	9.9	2.4

Source: Proceedings of the Euro-Construct Conference 1987)

The UK construction industry is the fourth largest in Europe, representing over 10% of the total output of work. It is surpassed only by Germany (30%), France (16%) and Italy (16%) (Table 1.2). New work typically accounts for 60% of the output, with repairs and maintenance representing the remainder. In the early 1970s public building and works accounted for over 50% of the construction industry's workload. By the early 1990s, this had fallen to less than 25%. In terms of new projects, public buildings currently account for 12%, with 88% being undertaken in the private sector. The public sector's decline is a result of the reduction in the overall size of the sector through privatization coupled with the pressure on government to control public expenditure.

Sectors

In 1970 the total value of construction output in the UK was £5.36bn (£30.99bn at 1985 prices). This increased to £22.05bn (£27.83bn) by 1980 and £44.70bn (£35.02bn) by 1990. The recession of the early 1990s saw considerable reductions in output in the succeeding years. Table 1.3 shows the percentage distribution of this work among the main sectors. The public sector share has reduced and is likely to continue to do so for the remainder of this century. For example, there are likely to be consolidation and improvements in education and health building projects rather than major new-build programmes.

The most promising growth area in the public sector for the 1990s is in the civil engineering infrastructure, with road and rail networks, sewerage schemes, telecommunications, and modifications to existing power stations. Little new local authority housing is likely to be constructed, with social housing being provided through housing associations. The demand for new private housing is expected to be revived towards the middle of the decade. Industrial development projects follow the fortunes of manufacturing industry and only a modest growth in this area is expected. There is, because of many of the recent new office buildings, an over-supply of commercial office space. An increase in business

Table 1.3   Sector share of the construction market (%).

	1970	1980	1990	2000 (forecast)
New housing				
Public	12.71	7.75	2.00	3.00
Private	13.46	11.74	12.21	17.00
Other new				
Public	23.74	15.96	12.05	17.00
Private				
Industrial	11.40	12.74	10.81	12.00
Commercial	10.28	11.01	21.44	9.00
Repair and maintenance				
Housing	12.52	20.31	22.10	22.00
Public	11.03	13.24	10.40	10.00
Private	4.86	7.25	8.99	10.00
Totals	100.00	100.00	100.00	100.00

Source: Housing and Construction Statistics and NEDO)

confidence is likely to result in upgrading and modernization rather than new-build projects. Retail developments offer some expectation of growth, particularly for those areas concerned with leisure activities.

The repair and maintenance sector will continue to remain a major component in the future. Building stock has often been poorly cared for in the past, and there is now a growing belief among some clients that such major capital assets need to be better managed in the future.

Employment

In 1990, the workforce in employment, which included those on government training programmes, was about 26.2m or 47% of the population. Unemployment stood at 1.5m, but by 1993 this had doubled to almost 3m. Self-employment, a culture that developed towards the end of the 1980s, increased to 3.25m. These trends were replicated in the construction industry. The construction industry directly employs about 1.7m people, most of whom are male. There are also a large number of others employed indirectly with materials and component manufacturers, plant and vehicles. A whole range of secondary employment also relies upon a prosperous construction industry.

The industry therefore typically employs about 4.5% of the employed labour force. The figure can fluctuate widely and may be as much as 25% higher in times of boom in the industry. Within these figures there are about 700 000 operatives, although owing to the cultural change in the nature of employment this figure declined steadily throughout the 1980s. At the same time the number of self-employed more than doubled to over 700 000. There are in addition a further 360 000 employed in administrative, professional, technical and clerical occupations. The number of people employed in the repair and maintenance sectors alone is greater than that of agriculture, coalmining, shipbuilding and many of the other traditional industries.

The development process

The life cycle of a construction project can be separated into five stages (Table 1.4). These are not discrete functions, and the different activities overlap at the various stages of the project's life cycle. The emphasis should be on securing developments that best satisfy all of these objectives rather than relying upon an appraisal of the initial expectations alone.

The inception stage

This is the stage in which the client will be determining the objectives for the project. A majority of projects arise from long planning programmes, in which the clients or promoters are considering the scheme as a part of the overall objectives of their own organization. The better-informed clients, who are often those involved in frequent capital development, usually have realistic expectations of what can and cannot be achieved in terms of time, cost and quality.

The type of project will often determine whom the client or promoter appoints as designer. For example, on engineering projects, civil engineering consultants are the most likely choice. On building projects the architect has traditionally been the first point of contact with the client. These are traditions that die hard. However, on smaller projects and schemes of refurbishment the building surveyor is being increasingly used as the client's main advisor. As the different combinations of design and build or management contracting are employed,

Table 1.4 The development cycle.

Stage	Phase	Typical time duration (years)
Inception	Brief Feasibility Viability	1
Design	Outline proposals Sketch design Detail design Contract documentation Procurement	1
Construction	Project planning Installation Commissioning	3
In use	Maintenance Repair Modification	80
Demolition	Replacement	—

Table 1.5 Characteristic times involved in the various stages of the development cycle.

	Inception stage (years)	Design stage (years)	Construction stage (years)
Public sector			
Housing	1–4	1–3	1–4
Health	1–5	0.5–4	0.5–5
Education	1–4	0.5–3	0.5–2.5
Other large buildings (law courts, civic buildings, etc.)	1–7	1–3	1.5–2.5
Other small buildings (libraries, etc.)	0.5–3	0.5–2	0.5–1.5
Roads and harbours	1.5–10	1–4	0.5–3
Water and sewerage	1–4	0.5–3	0.5–1.5
Private sector			
Housing	0.5–6	0.5–4	0.5–1.5
Industrial	0.5–2	0.5–2.5	0.5–2
Commercial	1–10	1–4	0.5–3

(*Source:* NEDO)

clients sometimes appoint the construction firm direct, choose an alternative consultant as a main partner in the venture or appoint a project manager in overall charge of the scheme. Table 1.5 indicates the typical time span of activities for different categories of projects. The long lead-in times at inception for some projects reflect the need to obtain the required finance and planning approvals.

Client's requirements

It is necessary in the first instance to determine the client's main requirements. Clients of the construction industry are wide and diverse, each with their own particular needs and desires regarding their project. The wide variety of contractual procedures that are now available reflect this fact. In principle, the client's requirements can be related to the following four main factors:

- *time*
  length of time for the design
  start and length of the contract period
  completion by the date stated in the contract
  time certainty.

- *cost*
  initial cost and relationship to tender sums and final cost
  value for money
  whole life-cycle cost approach.

- *performance*
  design in terms of function and appearance

construction reliability and performance
no latent defects; trouble-free guarantees
low-time and low-cost maintenance.

● *management*
clear allocation of responsibilities
accountability, particularly in the public sector
clear evaluation of risks.

The feasibility phase seeks to determine whether the project is capable of execution in terms of its physical complexities, planning requirements and economics. The available site, for example, may be impractical because of its size or shape, or the ground conditions may make the proposed structure too costly. Planning authorities may refuse permission for the specific type of project, or impose restrictions that limit its overall viability in terms of, for example, the return on the capital invested. Schemes may be feasible but may not always be viable.

The design stage

The alternative options and advantages of choosing either a separate designer and contractor or design and build are well documented. In either case it is first necessary to prepare schematic outline proposals for approval, prior to a detailed design. These proposals will need to be accepted by the client in terms of the requirements that have been outlined in the brief; by the relevant planning authorities, for their permission; and in terms of an outline budget by way of an initial cost plan. As the scheme evolves and receives its various approvals, a number of different specialist consultants will be employed. Some of these may be public relations consultants, particularly where the scheme is a sensitive one, such as a new road or building in the green belt, or where a project is proposed that is out of character with the locality.

The detailed design will follow when all the previous activities have been agreed and approved by the client. Different solutions to spatial and other design problems will be considered and some of these are likely to have a knock-on effect on aspects of the project that have already been agreed. Each alternative solution will need to be costed to ensure that the cost plan remains on target; where the alternatives significantly affect the client's proposals, this will need to be communicated for agreement. It will probably be necessary during this stage to involve the firms that supply and install the necessary specialist equipment.

The documentation that is required for tendering purposes will also be prepared at this time and concurrently as the design develops. When the project is approaching the tender stage, the different firms that may be interested in constructing the project should be invited to tender. Upon receipt of the documentation the contractors enter their estimating phase, since the awarding of the works of construction is most frequently done through some form of price competition.

The construction stage

This stage commences when the contractor commences the work on site. It is often referred to as the post-contract period, since it commences once the contract for the construction of the project has been signed and work has started on site. If the project is on a design and build arrangement or a system of fast-track procurement, then this stage may start before the design is finalized and will then run concurrently with it. Contractors are critical of the traditional arrangements as they are frequently required to price the works, which although assumed to be fully designed are in reality not so.

Throughout this stage formal instruction orders are given to the contractor for changes in the design, valuations are prepared and agreed for interim payment certificates, and final accounts are agreed. Contractual disputes may arise: all too often they are due to misunderstandings or incorrect information. The contractor is also sometimes over-ambitious and enters into legal agreements that become impossible to fulfil. This creates grounds for damages on the part of the client. Project completion times can be anything from a few months up to ten years or more. Upon completion, responsibility for the project is formerly signed over to the client, and the project enters its third stage in its life cycle.

The in-use phase

This is the longest stage of the project. The immediate aims of the client should now be satisfied, and the project can be used for the purpose of its design. During this stage occasions will occur when maintenance will be required. Even so-called 'maintenance-free construction' requires some sort of attention. The correct design, selection of materials, proper methods of construction and the correct use of components will help to reduce maintenance problems. A sound understanding, based upon feedback from project appraisals in practice, will help to reduce the possible defects that can arise in the future. Defects are often costly and inconvenient and minor problems can sometimes require a large amount of remedial work to rectify, out of all proportion to the actual problem.

Many projects have only a limited life expectancy before some form of refurbishment or modernisation becomes necessary. The introduction of new technologies also makes previously worthwhile components obsolete. City centre retail outlets, for example, have a relatively short life expectancy, before some form of extensive refitting is required. While the shell of buildings may have a relatively long life of up to 100 years, and some do last for much longer, their respective components wear out and need frequent replacement. Obsolescence may also be a factor to consider in respect of component replacement.

Demolition

The final stage in a project's life is its eventual demolition, disposal and the possible recommencement of the life cycle on the same site. Demolition becomes necessary through decay and obsolescence and when no further use can be made of the building or structure. Some buildings are destroyed by fire, vandalism or explosion, and may become dangerous structures that require demolition as the only sensible course of action. There are relatively few projects that last forever and become historic monuments, as the style of life and the needs of space are constantly evolving to meet new challenges.

Some notable projects become listed buildings. The Secretary of State for the Environment has powers under the planning Acts to compile lists of buildings that are of special historic interest. It then becomes an offence to demolish, alter or extend a listed building in any way that would affect its character as a building under this regulation. Where non-listed buildings are thought to have special historic or architectural interest than a planning authority may serve a building preservation notice upon the owner.

Contractors

At the start of the 1990s there were estimated to be over 210 000 construction firms operating in Britain, compared with 70 000 at the start of the 1970s. They ranged from sole proprietors to large multinational conglomerates employing several thousand people in their workforce. Over the same period the number of larger contractors – those employing over 1200 people – reduced from about 80 to less than 50. Construction firms can be categorized in many different ways, including the type of work that they undertake, the number of employees, turnover and location. Of the total number of construction firms almost 40% describe themselves as general builders.

There have been two major developments in construction firms over the past decade. The first has been the shift away from large general contractors to smaller specialist firms of subcontractors, with the main contractor taking on more of a management role. This had its origins in the recession in the industry in the early 1970s, and was given a further impetus during the 1980s with advances in technology, processes and specialization. The second development has been the change in the contractor's role, from an emphasis on construction alone towards an integrated single-point responsibility, such as design and build.

Employers

There are two main contracting employers' organizations in the construction industry: the Building Employers Confederation (BEC) and the Federation of Civil Engineering Contractors (FCEC). A third group, the Export Group for the Constructional Industries, includes those firms with an extensive or aspiring overseas interest. The membership of BEC is about 8000 and that of FCEC is 300. The size of firm in the FCEC is, on average, much larger. There are other subsidiary groupings, such as the House-Builders Federation (HBF) and the National Contractors Group (NCG).

The NCG consists of the largest companies who are in membership with the BEC, and includes about 75 firms. The NCG Code of Practice includes the following:

- Observe recognized standards of good practice.

- Use every endeavour to:
  complete contracts on time and within cost limits;
  fulfil obligations under the contract;
  help to promote the client's understanding of the contract;
  place emphasis upon achieving quality.

- Comply with the recognised standards of safety for both the public and the workforce

In 1993 the Chartered Institute of Building (CIOB) introduced its Chartered Building Company Scheme, in an attempt to improve the generally poor image of the construction industry. Any attempts to improve the image and reality must be welcomed. While the scheme is not a guarantee of workmanship or a warranty against defects it does require the company to be managed by professionally qualified people. A code of conduct ensures that participating companies discharge their duties to their clients and employees and engage only competent specialist subcontractors.

Unions

The number of trade unions in the construction industry has declined over the past decade as a result of amalgamations and mergers. There are now approximately ten different trade unions for construction workers. By far the largest number of construction workers are registered with the Union of Construction, Allied Trades and Technicians (UCATT). In the late 1980s the membership was 255 000. The membership in construction is relatively low. In 1980 it represented only 30% of all construction workers, but by the 1990s this had declined still further to about 20%. This was due to the growth in self-employment and the casual nature of employment. Despite efforts to attract a greater number of self-employed workers there has been relatively little success. Unless this trend is reversed their position will become further weakened. This can cause a loss of control of industrial relations by firms in construction and will be compounded by the wider spread of subcontracting.

Membership of unions by professional staff in the private sector is unusual. Individual surveyors attempt to negotiate their own salaries and conditions of service within the broad parameters that are set by a firm. Differences in practice between individual firms vary greatly, but reflect the supply and demand of surveyors, the responsibilities offered and the opportunities provided. In the public sector, at all levels, it is more usual for surveyors to be members of the recognized trade union.

Other organizations

There are numerous organizations representing different activities and interest groups both generally, and specifically for the construction industry. They include, for example, the following:

- *Materials and manufacturers*
  Agrément Board
  British Cement Association (BCA)
  British Standards Institution (BSI)
  British Steel Corporation (Constrado)
  Timber Research and Development Association (TRADA)

- *Research*
  Building Research Establishment (BRE)
  Construction Industry Research and Information Association (CIRIA)
  European Network of Building Research Institutes (ENBRI)
  International Council for Building Research Studies and Documentation (CIB)

- *Education and training*
  Construction Industry Council (CIC)
  Construction Industry Standing Conference (CISC)
  Construction Industry Training Board (CITB)
  National Council for Vocational Qualifications (NCVQ)

- *Costs and contracts*
  British Property Federation (BPF)
  Building Cost Information Service (BCIS)
  Building Maintenance Information (BMI)
  Joint Contracts Tribunal (JCT)
  National Joint Consultative Committee (NJCC)
  Standing Joint Committee for the Standard Method of Measurement.

The professions

The professions have been one of the fastest-growing sectors of the occupational structure in Britain. At the turn of the century they represented about 4% of the employed population. In the early 1970s this had risen to over 11%, and the trend in their growth has accelerated as the service sectors increased their importance during the 1980s, and manufacturing either became more mechanized or generally declined. The temporary lull in the expansion of the professions due to the recession of the 1990s has caused much discussion on their benefits to society. A similar trend for comparable groups is evident in all Western capitalist societies. Several reasons are given for the rapid growth of the professions, such as an increasing complexity of commerce and industry, the need for more scientific and technical knowledge, and a desire for greater accountability.

However, the professions are vulnerable in this age of computer development and advancement. Some of their repetitive functions are affected by the impact of computer mechanization, and it will become possible to access their specialist knowledge through the advent of expert systems. Easy access to telecommunications will allow the professions' work to be carried out in the most economical parts of the world. Non-construction professions are now competing for a share in the workload, and the professional services that today might not be thought necessary will become commonplace during the next century.

The built environment professions

The built environment professions in the UK are many and varied, and represent a distinctiveness for the industry and a matter for much debate. There are almost 280 000 (members and students) among the seven main chartered professional bodies that work in the construction industry (Table 1.6). In addition to the chartered professions there are over 20 different non-chartered bodies, which include the Institution of Surveyors, Valuers and Auctioneers and the Architects and Surveyors Institute.

The Royal Institution of Chartered Surveyors (RICS) is the largest professional body in construction, with a membership of over 85 000. It is sometimes argued that the difficulties that arise in the industry are due, at least in part, to the many different professional groups that are involved. Others argue that the services that the construction industry provides have now become so specialized that one or two different professional groups would be inadequate to cope with the complexities of the construction process. In this respect, the UK is out of step with the rest of the world. However, there is no standardization of practice, and considerable differences exist even across the different countries of Western Europe.

There are also wide cultural considerations to be taken into account in any comparison between the construction industry professions in the UK and those in other parts of the world, notably Europe, the USA and Japan. Historically, practices developed differently. In much of the rest of the world, architects and engineers dominate the construction industry. The various professional disciplines in Britain are not mirrored elsewhere, other than in Commonwealth and ex-Commonwealth countries. The role of the professional bodies also varies. In the UK a professional qualification is one by which to practise. In Europe a professional body is more of an exclusive club, to which relatively few of those engaged in practice belong. In the USA there is the emerging discipline of construction management alongside those of the architect and the engineer.

Table 1.6   Professional body membership.

	Members	Totals
The Royal Institution of Chartered Surveyors	64 724	86 830
Institution of Civil Engineers	49 665	75 976
Chartered Institute of Building	10 246	31 654
Royal Institute of British Architects	28 500	31 000
Institution of Structural Engineers	12 008	21 520
Royal Town Planning Institute	12 500	16 500
Chartered Institute of Building Services Engineers	8 814	15 250
Totals	186 457	278 730

(*Source:* Professional institutions)

Architects, engineers and surveyors

The architect is the designer of the building project and has the difficulty of translating the building owner's ideas into an acceptable design and then into working drawings. The profession of architect is at present a registered title, through an Act of Parliament. It is the only registered profession in the construction industry in the UK, although this privilege is likely to cease in the near future. While this Act currently prohibits anyone that is not registered from calling themselves an architect, it does not prohibit them from doing an architect's work or from using a title such as 'architectural engineer'. In this respect registration differs from that of a doctor or dentist. Elsewhere around the world surveyors as well as architects may also need to be registered.

The architect relies upon the quantity surveyor's efficiency and cooperation, together with a mutual understanding of each other's problems. The advice of the quantity surveyor on matters of cost and financial matters is important throughout the development process. In addition to the design the architect will deal with all the necessary approvals that are required under the various statutory regulations and controls, such as planning permission and Building Regulation approval.

One of the architect's difficulties is to combine the artistic sense of design with the practical aspects of construction. Other than in the more common forms of buildings, the architect is unable to carry out the necessary calculations in terms of either structural components or engineering services. This type of work requires a different temperament from that of an artist. Consulting engineers specializing in these respective aspects of construction are therefore employed to ensure that the project is structurally sound and environmentally effective. The quantity surveyor works closely with these engineers and in some circumstances is required to prepare separate documentation for their work. In addition the quantity surveyor ably fulfils the cost role on civil and heavy engineering projects.

Quantity surveyors also undertake projects with other kinds of surveyors. They work with building surveyors on smaller projects, providing the required cost and contractual information in much the same way as is offered to architects on major schemes. The building surveyor's expertise is concerned with monitoring the condition of building projects, as owners have realized that their buildings are their major investment. The quantity surveyor will also work with the general practice surveyor, whose expertise relates to investment, and as such is often employed by development companies.

The quantity surveying profession

In 1992 there were 32 516 members of the QS Division of the RICS (Fellows 9380 and Professional Associates 23 136). The Division has been growing at a rate of about 3% per annum. The number of entrants to quantity surveying degree courses during the last decade rose faster than for any other construction discipline. This reflects its popularity as a career and the opportunities that it provides.

There has been a shift in employment patterns from the public sector towards contracting (Table 1.7). While the profession has weathered the economic storm

Table 1.7   Employment patterns of quantity surveyors (%).

	1983	1993
Private practice	53	52
Public service	22	15
Contracting	17	21
Commercial	6	10
Education	2	2

(*Source:* RICS)

of the early 1990s better than many other professions, it has nevertheless suffered in terms of employment and remuneration.

In comparison with other professions, such as engineering, accountancy and law, quantity surveying private practices are small. The largest QS practice has about 1500 staff worldwide, and this is exceptional. By comparison the largest British engineering consultancy has almost 4000 staff worldwide. On an international scale some of the larger accountancy and law firms employ in excess of 5000 staff. At the other extreme the smaller practices, with fewer than ten staff, account for between 50% and 60% of QS private practices. There has been a trend during the past decade to develop integrated practices employing architects, engineers and surveyors.

The future of the built environment professions

The built environment professional bodies have grown steadily both in membership and in number throughout this century. The number of different professional bodies has continued to increase in spite of the mergers that have taken place. It can be argued that there are too many bodies working in the construction industry in the UK. Their future is influenced by:

- the effects of the Single European Market;
- the diversification and blurring of professional boundaries, often including non-built environment professions such as those involved with the law and finance;
- their role as learned societies;
- the education structure of courses in the built environment;
- the pressure groups both inside and outside the construction industry;
- the desire in some quarters for the formation of a single construction institute, to unify all professionals in the construction industry.

The future of the construction industry

Some of the major issues that will influence the structure of the construction industry up to the beginning of the twenty-first century are as follows:

- response to the booms and slumps in the economy and the stop-go policies of government;

- traditional separation of the design and production of buildings;
- fragmentation of the professions in Britain;
- quality of its products;
- time required from inception to completion;
- overall costs involved;
- dearth of research and development (see Chapter 15);
- increased emphasis upon environmental considerations;
- labour relations, particularly as the economy moves out of recession;
- competition from Europe, USA, Japan;
- new international markets;
- Third World developments.

Bibliography

Bagenal, L. and Meades, J. *The Illustrated Atlas of the World's Great Buildings*. Salamander Books, 1987.

Brandon, P.S. (ed.) *Quantity Surveying Techniques: New Directions*. Blackwell Scientific Publications, 1992.

Centre for Strategic Studies in Construction *Building Britain 2000*. University of Reading, 1988.

Centre for Strategic Studies in Construction *UK Construction Prospects 2001*. University of Reading, 1990.

CIOB *Building for Industry and Commerce: A Client's Guide*. Chartered Institute of Building, 1980.

Davis, Langdon and Everest *QS 2000: The Future of the Chartered Quantity Surveyor*. The Royal Institution of Chartered Surveyors, 1991.

Harper, D.R. *Building: The Process and the Product*. Construction Press, 1978.

Harvey, R.C. and Ashworth, A. *The Construction Industry of Great Britain*. Newnes, 1993.

Hillebrandt, P.M. *Analysis of the British Construction Industry*. Macmillan, 1984.

Housing and Construction Statistics 1980–1990. HMSO, 1991.

NEDO *Construction Forecasts 1991, 1992, 1993*. National Economic Development Office, 1991.

RICS *Surveying in the Eighties*. The Royal Institution of Chartered Surveyors, 1980.

RICS *The Future Role of the Chartered Quantity Surveyor*. The Royal Institution of Chartered Surveyors, 1983.

UK Construction: The Industry, Its Markets and the Prospects for the 1990s. Economist Intelligence Unit, 1990.

Chapter 2

The work of the quantity surveyor

Introduction

The work and services provided by the quantity surveyor today may be described as the financial management of the project, whether it be on behalf of the client or the contractor. The term 'quantity surveying' does not now reflect the services that are provided, as these have been extended to cover what might be more appropriately termed 'project cost management'. Much discussion has taken place regarding a change in name but the prevailing view rests upon the assumption of the familiarity of the name of the quantity surveyor. The profession is not unique in having this problem.

In 1971 the RICS published a report entitled *The Future Role of the Quantity Surveyor*, which defined the quantity surveyor as:

> 'ensuring that the resources of the construction industry are utilized to the best advantage of society by providing, *inter alia*, the financial management for projects and a cost consultancy service to the client and designer during the whole construction process'.

The report emphasized that the distinctive competence or skills of the quantity surveyor were, in the wider aspects of the construction industry, associated with measurement and valuation. This provides the basis for the proper cost management of the construction project in the context of forecasting, analysing, planning, controlling and accounting.

In 1983, owing to the rapid developments in the profession, the RICS considered that a new report should be prepared that would explore further the work of quantity surveyors and also try to assess their future potential and directions. This report, *The Future Role of the Chartered Quantity Surveyor*, identified a wide range of skills and expertise, and suggested an even greater expansion of services and, in some cases, in industries other than construction.

In 1991, a further report was commissioned by the Quantity Surveyors Division of the RICS, entitled *QS 2000 – The Future Role of the Chartered Quantity Surveyor*. This report in its analysis pointed towards a key message of change by outlining the threats and opportunities that were facing the quantity surveying profession. These changes included:

- *changes in the markets*: outlining the previous performance and trends in workloads across the different sectors and the importance of the changing international scene, particularly the challenges from the European Union.

- *changes in the construction industry*: through the changing nature of contracting, an emphasis upon management of construction, the comparison with other countries abroad and the competition being faced from non-construction professionals;

- *changing client needs*: with an emphasis in terms of the value added to the client's business, they want purchasable design, procurement and management of construction;

- *changes in the profession*: noting employment patterns, the growth in graduate members, the impact of fee competition, the ways in which the quantity surveyor is now appointed, and changes in their role and practice with changing attitudes and horizons.

The work of quantity surveyors can therefore be summarized briefly as follows:

- preliminary cost advice;
- cost planning including investment appraisal, life-cycle costing and value analysis;
- procurement and tendering procedures;
- contract documentation;
- evaluation of tenders;
- cash-flow forecasting, financial reporting and interim payments;
- final accounting and the settlement of contractual disputes;
- cost advice during use by the client.
- project management
- specialist services.

Traditionally, certainly during the early part of this century, quantity surveyors were employed as preparers of bills of quantities for building projects. Their role was constrained to a limited but important part of the development process. This role was quickly extended to include the preparation of valuations for interim certificates and the agreement of the final account with the contractor.

During the 1960s the quantity surveyor's role was enlarged to include design cost planning, which many believed provided the solution to abortive bills of quantities, while offering the client some form of value for money and cost effectiveness. In more recent times greater emphasis has been placed on the need to examine construction costs in terms of their life-cycle rather than solely in terms of initial costs.

Some quantity surveyors have seen the application and relevance of their work to problems of costs-in-use associated with the client's occupation of buildings. In order to complete the picture it is now necessary for the quantity surveyor to become fully involved at the outset of a project's development. Although lip service has been paid to this in the past, the designer has often completed this stage of the development process by relying only upon a limited input from the quantity surveyor. It is during this stage that the type and size of the project are largely determined, and these two factors alone commit a considerable proportion of the project's total expenditure.

It would of course be unfair, and incorrect, to presume that the project's development from now on only requires the designer and the surveyor to 'tinker' with the costs. In the context of a tender sum the whole process can only be considered as a period of cost refinement, however important and valuable this may be. Quantity surveyors must therefore make a proper and sizeable contribution during the process of strategic planning. They must also become familiar with the special needs of the client in order that they can properly evaluate the options that are under consideration.

The management of project costs is shown graphically in Fig. 2.1.

Client demand

In 1984 the RICS conducted an extensive survey on quantity surveying practice and client demand and published a report under that title. The study was undertaken to establish the availability of the range of services offered by the profession, and the requirements of the industry's clients.

The study of the profession showed that quantity surveyors were already established in a wide diversity of work. Over half the profession were employed in private practice, with bills of quantities still representing a significant part of the workload, although this had diminished in importance in recent years. An interesting statistic arising from the study is that 42% of the practices had been formed since 1970, and that a number of practices now offered specialist services such as arbitration, loss adjusting and construction management etc. UK work accounted for 93% of the total workload and of this 36% was represented by public sector projects.

Clients of the quantity surveyor have varying needs. In some cases they will be unaware of the services provided while on other occasions they may require advice of a rather specialized nature. They do, however, require complete and comprehensive advice on which they can rely. Their concern is not so much with the techniques that the surveyor may use as with the value of the benefits that can be provided. They may require the surveyor to evaluate the possible options that are available. They will certainly require a recommendation of the best course of action from their own viewpoint. It is vitally important therefore that the surveyor fully understands the needs and aspirations of the client at a very early stage in the construction process.

The quantity surveyor's proven expertise is primarily on construction costs. More clients are also now concerned with the implications of matters of time, and the interaction of time and cost. The quantity surveyor is also required to bear in mind the final value of the project in addition to the prediction and accounting of its costs. Clients also require regular contact with the quantity surveyor and advice on all issues of importance to them. This advice, to be of any use, must be accurate at all stages and both impartial and independent in character.

Clients are sometimes dissatisfied with the services provided by the quantity surveyor. Some of this dissatisfaction arises from discontent with the construction industry and its processes in general. Some would argue that the industry is archaic in its attitudes and in need of fundamental changes. Surveyors sometimes fail to appreciate and identify with the client's objectives and constraints. The services provided often take too long to perform, and are untimely and

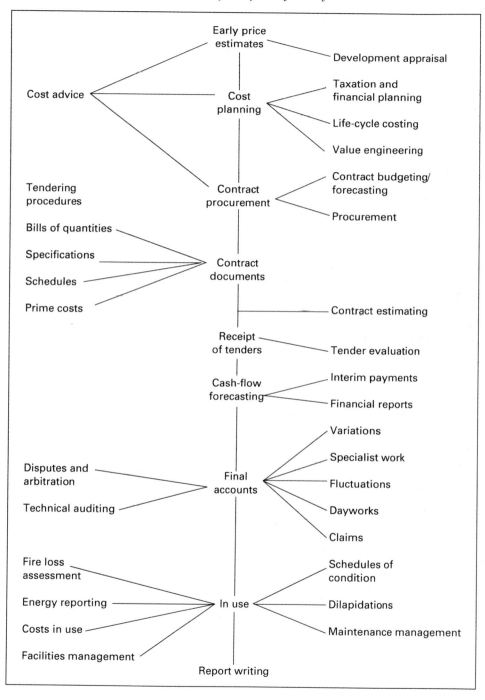

Fig. 2.1 The work of the quantity surveyor.

inappropriate to the client's needs. Some clients feel that quantity surveyors are not sufficiently commercially minded. Surveyors may believe that this requirement can at times be in conflict with their professional status.

Criticism has also been expressed of the quantity surveyor's primary role concerning building costs. The quality of the advice and cost control was sometimes inaccurate and inconsistent, and cost forecasting was not good. This criticism may have arisen because the quantity surveyor failed to appreciate the client's needs, or may be due to the surveyor's specific inexperience. It may also in part relate to a lack of authority or control on the part of the surveyor. In other circumstances, clients may unreasonably expect a level of forecasting that is impossible to achieve in practice. The profession must, however, be continually striving to improve its performance in this matter through research and development.

Professional conduct

The concept of a professional person is continually evolving but at present it generally refers to someone who offers competence and integrity of service coupled with a skilled intellectual technique. The relationship between such persons and their clients is one of mutual trust. Although in recent years there has been a blurring of professional and business activities, the distinction between professional and commercial attitudes remains.

Membership of any organization requires compliance with its rules. Members of the RICS must comply with *the Rules of Conduct, Disciplinary Powers and Procedures* which are published by their institution. These rules have been the subject of increasing discussion, negotiation and amendment in recent years. They have been devised to protect the interests of the profession's clients and to help maintain its status and image in society. They seek to set an acceptable standard of practice based upon the attitudes and support of the membership of the Institution, the requirements of statute, and public opinion. In more recent years there has been an easing of some of the restrictive codes, allowing members more scope to develop their expertise and services within a more competitive and challenging business environment. Matters covered in the Rules include conflict of interest, financial interest, status and designations, fee quotations, advertising and publicity.

Chartered quantity surveyors must carry out their work in a manner befitting that of their profession and the Institution. Several courses of action are open to the Institution if a member contravenes one of its by-laws. It can:

- admonish the member
- require the member to refrain from the contravening conduct
- reprimand the member
- suspend the member
- expel the member.

These disciplinary powers exercised by a Professional Practice Committee, a Disciplinary Board and an Appeals Board.

Quantity surveyors' workload figures

The RICS collect and analyse data on the workload of quantity surveying practices. Although the information gleaned is only of a general nature it does nevertheless provide some guidance as to the trend or pattern in workload. The results of the survey are published in the *Building Economist*. The information is based upon a sample number of firms being asked the single question of whether the practice is more or less busy now than it was three months ago. From this a histogram is prepared based upon percentages. In addition, a brief commentary is provided on the salient features of the results. Any other surveying practice can then gauge its own performance against such a national or regional trend.

Skills and knowledge base of the quantity surveyor

The RICS in a report published in 1992 entitled *The Core Skills and Knowledge base of the Quantity Surveyor* examined the needs of quantity surveyors in respect of their education and training and continuing professional development in the face of change and uncertainty in the industry and practice. The report identified a range of skills that the profession would need to continue to develop if it wished to retain its focal position within the construction industry. The report identified a knowledge base that included:

- construction technology
  measurement rules and conventions
  construction economics
  financial management
  business administration
  construction law

and a skill base that included

- management
  documentation
  analysis
  appraisal
  quantification
  synthesis
  communication.

The report also set out to identify the present and future markets for this knowledge and skills and the constraints that might inhibit quantity surveyors from achieving their full potential. The report developed the strength and ideas from earlier reports published by the RICS and the issues identified in the Quantity Surveyors Division report *QS 2000*, by examining the key trends in the demand for construction activity and the needs for professional services. The report also made reference to the wider opportunities beyond the horizon of construction where the knowledge and skill base could be applied.

In analysing the knowledge base and accepting that this was incremental and on a need-to-know basis, the report identified four key areas of:

● *technology*: relating to the process and the product
● *information*: sources and information management
● *cultural*: organizational and legal
● *economic*: business and finance.

Differences between skills and techniques were also identified. Quantity surveyors use many different techniques, some of which are mechanical in their approach. Skills occur in respect of the levels of ability required to apply these techniques expertly. These different skills can be assimilated with the knowledge base through education, training and practice. While there is general agreement on the knowledge and skills base that is required, different surveyors will place different emphases upon their importance. The report concludes with a forecast of the future importance of the different core skills and knowledge that might be expected in five years time.

Types of project

Quantity surveyors are largely concerned with the following four main areas of work.

Building work

The employment of the quantity surveyor on building projects today is largely taken for granted. The introduction of new forms of contract and changes in procedure has altered the way in which quantity surveyors carry out their work, but they occupy a much more influential position than in the past, particularly when introduced into the procurement process at an early stage. Quantity surveyors are essentially cost experts whose responsibility it is to advise the client upon the cost implication of design decisions and, where possible, to influence the control of costs.

Great importance is placed upon the control of costs on most projects. However, while quantity surveyors may have some influence in controlling costs the true controller of costs is the person (client or architect) that orders the work that generates the cost. If the quantity surveyor is properly integrated into the design team, with a clear brief for the management of costs, which all design consultants are required to adhere to, it is possible to prevent costs getting out of hand.

Building engineering services

Although this work is very much part of the building project, it has become a specialist function for the quantity surveyor. Traditionally, much of this work has been included in bills of quantities as prime cost sums. It was largely presented in this way for three reasons: services engineers often failed to provide the appropriate details for measurement; quantity surveyors had been unaccustomed to measuring in this way; and contractors preferred to offer quotations on the basis of drawing and specifications only. The more enlightened clients will now usually favour the use of bills for engineering services, although there is still some resistance to them from services consultants.

It is, however, also accepted that to provide a rigorous cost control function for only part of a building project is not entirely satisfactory. The services engineering work is in addition becoming a more extensive and costly part of the project, emphasizing the need to apply improved and more appropriate methods of cost control. Quantity surveyors who are employed in this discipline of the profession have had to become more conversant with engineering services technology and terminology, and to interpret engineering drawings correctly.

Civil engineering

It is impossible to define the line between building and civil engineering works accurately. The nature of civil engineering projects often requires a design solution of physical and geological problems. The scope, size and extent of the works to be performed are frequently considerable. These problems will dictate the cost of the solution, and the engineer must be able to provide an acceptable one within the confines of a predetermined budget in the same way that buildings can be cost-planned within cost limits. They can involve a large amount of uncertainty, owing to the nature of ground conditions and temporary works, which can sometimes represent a major part of the project.

Civil engineering projects use a different method of measurement (Civil Engineering Standard Method of Measurement or Method of Measurement for Roads and Bridgeworks) and different conditions of contract (ICE conditions). This is due to the different nature of civil engineering works and the different approach that contractors have towards it. The work is said to be more method-oriented than building works, with a much more intensive use of mechanical plant and temporary works. The bills of quantities therefore comprise large quantities of comparatively few items. Generally at least parts of the work are of an unforeseeable nature, and civil engineering quantities are normally therefore approximate, with full re-measurement of the work actually done.

Quantity surveyors working in the civil engineering industry will provide similar services to those of their counterparts working on building projects. In addition to the methods of measurement and conditions of contract, the surveyor must also be conversant with the different working rule agreements, daywork rates and other matters of relevance, such as different methods of working. *Civil Engineering Procedure*, published by the Institution of Civil Engineers is a useful reference.

Quantity surveyors have now largely been accepted as members of the engineer's design team, and have been employed by civil engineering contractors since the turn of the century. The quantity surveyors' contribution is largely due to their expertise associated with construction costs in measurement analysis and valuation. Engineers now value the advice that the quantity surveyor is able to provide. The quantity surveyors' education and training furnish them with the necessary skills and expertise that only the specialist is able to provide.

Engineers themselves have shown a preference towards engineering design, and are generally content to allow someone else to look after the cost and financial aspects. The promoter has to know the likely cost of the project in advance of construction, and the quantity surveyor's experience can be usefully employed to provide the engineer with comparative costs of alternative solutions. After tender, the quantity surveyor may be asked to analyse and report on tenders and, after the contract is awarded, may be called upon to evaluate the final cost, including remeasurement, variations and assistance in claims or adjudication.

Heavy and industrial engineering

This work includes such areas as onshore and offshore oil and gas, petrochemicals, nuclear reprocessing and production facilities, process engineering, power stations, steel plants, and other similar industrial engineering complexes. Professional quantity surveyors have been involved in this type of work for a relatively short period of time, but because of changing circumstances within these industries and a greater emphasis on value for money, they are now being accepted by many clients and contractors as very valuable members of the project team. In an industry that employs a large number of specialists, the quantity surveyor, with his practical background, commercial sense, cost knowledge and legal understanding, has much to offer.

This work is generally classified as cost engineering and has traditionally been carried out by chemical engineers or the designers of the plants. Modern-day cost engineers are, however, more likely to have their roots in quantity surveying. The professional cost engineer is widely employed in the USA and many countries of Europe, and this is growth profession. There had not been any recognized form of contract or method of measurement until the RICS and the Association of Cost Engineers prepared their Standard Method of Measurement for Industrial Engineering Construction (SMMIEC), and standard forms such as those published by the Institution of Chemical Engineers were developed.

The basic methods employed are not very different from those used in other typical quantity surveying work. They are often, however, more numerically based and offer different forms of analysis which lend themselves to computerized measurement, billing and cost administration systems. Bills of approximate quantities are often produced from sketches and drawings provided by one of the design departments. Otherwise, performance specifications or schedules of rates or one of the forms of cost-reimbursable contract may be used. Alternatively take-offs are prepared for the purchase of materials only.

Essentially, the quantity surveyor employed on this type of work must be flexible enough to adapt to new methods of measurement, cost analysis and contract procedures. There is also a likelihood of being involved in a wider range of activities than for a quantity surveyor employed on building projects alone.

Types of work

Precontract cost control

One of the quantity surveyor's early duties on any construction project is to suggest to the client possible guide figures on the likely cost of his scheme.

Where this is acceptable to the client or where a figure can be agreed upon, an approximate estimate for the project is prepared using one of the methods described in Chapter 8. Throughout the design stage the quantity surveyor will be required to provide cost advice and cost comparisons of the different methods of construction that may be envisaged. Wherever possible the client should be persuaded, if this is necessary, to adopt a form of cost planning. The purpose of this is to attempt to control the design in terms of cost and also to provide some value-for-money objectives. This may be coupled with a form of value analysis. It is also likely to reduce the need for an addendum bill of quantities and the subsequent delay that this might cause to the project.

In addition it may be necessary to undertake some form of life-cycle costing in an attempt to take into account maintenance and running costs in addition to capital costs alone. A part of the pre-contract function will also be to prepare a budget for the client based upon their total capital cost expenditure, taking into account other costs in addition to construction costs.

Contract procurement

Some quantity surveyors have always been in the vanguard of contractual developments. The Economic Development Committee (EDC) report *Faster Building for Industry* stated that projects were traditionally often organized by default and that clients were unaware that they even had a choice. This criticism can be remedied by using the quantity surveyor's special expertise in procurement procedures, particularly when the quantity surveyor can be approached at the outset for guidance in this area.

There are a wide range of alternative procedures available (see Chapter 9) and a problem exists where their evaluation is based upon opinion or bias alone, resulting in different recommendations from different surveyors. All clients, in both the public and private sectors, are concerned with the trio of time, cost and quality. They may assess these differently, but as far as cost in concerned they are more interested in knowing the final cost with certainty then solely in choosing a lowest price tender. Hence they are interested in new forms of contractual arrangement that satisfy their needs better than the traditional processes.

Design and build

Something needs to be said about design and build types of contract as the services provided by the quantity surveyor will be different from those required when engaged in a more traditional form of client relationship. The quantity surveyor may be commissioned by the contractor along with other members of the 'design' team, the difference being that the fees will be paid by the contractor and the service offered may be less elaborate. Alternatively the client may choose to retain the services of an independent quantity surveyor to keep a watching brief on costs and payments to the contractor. The extent of the surveyor's involvement will depend upon the type and amount of advice required by the client.

Contract documentation

A major and important part of the quantity surveyor's workload is in the pre-paration of the contract documentation. This varies depending upon the nature, extent and size of the project, but can be summarized as follows:

- bill of quantities
- schedules of rates
- specification
- forms of contract.

This aspect of a quantity surveyor's work is dealt with in detail in Chapter 10, which also includes the steps necessary to obtain and evaluate tenders.

Contract administration

The quantity surveyor's skills in this field are perhaps some of the more neglected and underrated. The financial control of any construction project continues until the issue of the final certificate, and the method of control and the duties to be performed by the surveyor will depend upon the method used for contractor selection, the method used for price determination, and whether the control is required on behalf of the client or the contractor. The cost control aspects of a quantity surveyor's work are covered in detail in Chapter 8 and the post-contract procedures in Chapter 11.

Work for contractors

Whether a surveyor in private practice should undertake work for builders or not is a matter of policy to be decided.

The work required is usually in one of two categories – either measurement or pricing – and sometimes perhaps both. Measurement may be required where no quantities are supplied for tender, for the adjustment of variations on a contract or, in the case of schedule contracts, for the complete measurement of works as executed. Any of these is within the province of the quantity surveyor.

Pricing of estimates is a different matter. Quantity surveyors have the theo-retical knowledge of price analysis that they can apply in practice, but success in tendering in competition depends very largely on a detailed knowledge of current prices, which firms to approach for the best terms and the best service, and above all a decision on the risks to be taken. In the first two of these the contractors, constantly dealing with merchants and handling invoices, are obviously more expert, and the last, after all, is their affair. If A is entrusted with taking risks with B's money, there is a tendency to be more cautious than B might be, as A will be liable to be blamed for any loss.

Measurement is a question of fact, but pricing is opinion based on judgement and that knowledge that only the constant handling of labour and purchase of materials can give. To price estimates merely by taking rates from price books or other bills may, when carefully done, be suitable for approximate estimates; it does not give the accuracy necessary for pricing tenders, where quite a small margin one way may lose the job, or the other may seriously reduce or even eliminate the profit.

Disputes, litigation and arbitration

Quantity surveyors are often involved in disputes when differences have arisen between the parties on a project on which they are involved, and the matter has become the subject of litigation. They will be required to provide evidence and on occasions be required to act as a witness of fact. There is also scope for experienced quantity surveyors to act as expert witnesses or arbitrators. The role of the quantity surveyor in all of these matters is covered in Chapter 13.

Specialized services

Specialized services that quantity surveyors can provide for their clients include:

- technical auditing
- valuation for fire insurance
- fire loss assessment
- schedules of conditions and dilapidations
- adjoining owner matters
- capital and tax allowance advice
- facilities management
- land and acquisition advice
- development appraisal
- project management.

These services are described in detail in Chapter 14.

Representation of a contractor

It sometimes happens that a contractor needs to be represented by a quantity surveyor in a private practice. It may be that the contractor is short of staff or is too busy, or it may be that such representation will help in a contract difficult because of its specialized nature. There are some surveyors who make a special feature of such work for contractors and others who dislike it. In a country practice, where one is constantly dealing with local architects, it may put a surveyor in an awkward position through having to act as a quantity surveyor under one contract and independently for the contractor in another, both perhaps with the same architect. In a practice in London or a large provincial town, where surveyors generally go wider afield, there is not the same difficulty. If asked to act for a contractor in a contract in which the architect is one whom one is working with or might have to work with at another time, it would be a courtesy to mention this to the architect before accepting.

The role of contractor's surveyor is more fully developed in Chapter 4.

Reports

Although a surveyor may normally specialize in quantity surveying, in certain places it is inevitable that such work will be part of a general practice. In such cases the surveyor is fairly certain sometimes to have to prepare reports on buildings. If carrying out this type of work it is essential that the conditions of the surveyor's professional indemnity policy are carefully examined to ensure that the policy does not, as many do, specifically exclude structural surveys. Even though the survey may be carried out without remuneration there will still be a liability for negligence.

The essence of a successful report is that it should cover the points on which the client wants information, and not contain a lot of unwanted or unnecessary matters. A prospective buyer's chief concern is the building's stability and state of repair, which will be reflected in the bills for maintenance during the coming years. Certain alterations may be envisaged and the report will be required to cover this aspect. Other matters that may be required include electricity and drain tests, and these too will need to be covered or specifically excluded if not done.

Project management

Because of their logical form of training and their organizing ability some quantity surveyors have emerged as project managers. This is more fully considered in Chapter 12. Quantity surveyors may also provide general management services in connection with construction projects.

Research and development

Some of the larger practices have thought it appropriate to incorporate a research and development (R & D) section in their offices. Perhaps it is only the larger firms who are able to devote resources to this, but in a smaller way all practices will in some manner be considering regularly the development of the practice and its services. R & D is considered to be an important and integral part of industry, so it must also have its place in the professions, and more particularly in quantity surveying practices. The subject of R & D is covered in more detail in Chapter 15.

Overseas work

Opportunities exist overseas either for quantity surveyors to be employed by foreign governments or firms or for practices to work for overseas clients. In recent years the construction boom in parts of Africa, such as Nigeria, and in the Middle East, has provided excellent experience for quantity surveyors. Recent delegations from the RICS to both China and Japan will result in the profession's becoming more established in those countries.

An RICS report stated that quantity surveyors were now being employed in over 100 countries around the world. They have for a considerable period of time been employed in virtually all Commonwealth countries, and encouragement and assistance can be received from the Commonwealth Association of Surveying and Land Economy (CASLE). They are at work in several countries in mainland Europe, where they have established the Construction Economics European Committee. In the USA there is the American Institute of Construction

Economics. Several UK-based practices have formed associations or links with indigenous practices abroad.

The RICS prepared its publication of the Principles of Measurement (International) for Works of Construction, although many of the overseas countries have now developed their own methods of measurement.

Several practices have undertaken the costly enterprise of setting up abroad to provide an on-the-spot service, with all the problems, political and financial, that this involves. Opportunities abroad, like all opportunities, are there to be taken, and it must be up to the practitioner to make the decision; all that can be said is that there are many occasions when the failure to take up an opportunity has been regretted afterwards. The actual work differs only in the respect that professional ethics, traditional in the UK, do not necessarily apply abroad. In fact the reverse is often the case, and a quantity surveyor must be prepared to proceed in a much more competitive way, not only on the question of fees but in some of the short cuts he or she may be forced to take.

Future role of the quantity surveyor

Although the role of the quantity surveyor has changed considerably since the Second World War this is being overshadowed by the current changes that are occurring. The industrial revolution that took place at the beginning of the nineteenth century is currently being followed by a revolution in commercial activity. The ways in which management performs its duties are undergoing an even bigger change than that which affected the shop-floor worker. This is likely in the long term to create immense repercussions on a large scale within the profession.

There is now a need for flexibility in order to respond to new demands and opportunities, and to ensure independence from the other professions for their source work or information.

The industry's clients have become increasingly concerned about the services provided. They have examined traditional contractual procedures, have often found them lacking, and have sought to create new ones more appropriate to their needs. In general they have been looking for speedier results and improved value for money. Some caution does, however, need to be exercised. It is possible to devise a highly complex and sophisticated set of procedures, which would be extremely costly to administer and for which clients could generally not afford to pay. Any development must therefore be realistically tempered with the practicalities, otherwise it becomes nothing more than an ideology. Few clients will be able to adopt the maxim of Cheops who said when commissioning his pyramid, 'I don't care how much it costs or how long it takes'.

The present age is therefore one of cost-consciousness and desire for value for money. Indeed, a truly good architectural or engineering design will always take the cost implication into full account. This is a sensible approach and one that is likely to be applied even more rigorously in the future.

One thing, however, should be recognized as virtually certain, and that is the ever-increasing use and application of information technology to assist in the work of quantity surveyors, much of which is relevant to computer usage. Although at the moment the majority of this is in commonplace applications, current developments are more far-reaching. The quantity surveyor is having to cope with the impact of computer-aided design, and construction costs being evaluated and appraised in a completely different way. Computerization should not be looked at with concern, since it enhances the professional service while removing some of the tedium. There will also be some tasks of the profession that will not be capable of being carried out by computer. Someone of professional standing will be required to service and update the system; clients will still prefer to converse with people, and will still value professional opinions.

Bibliography

Amery, C. *et al.* 'Quantity surveying'. *Financial Times*, 4 April 1984
Ashworth, A. *Life Cycle Costing: Can It Really Work?* American Association of Cost Engineers, November 1989.
Ashworth, A. 'Loss adjusting and the quantity surveyor'. *QS Weekly*, 10 December 1981.
Bar-Hillel, M. 'Master of all trades? The case for the quantity surveyor as project manager'. *Chartered Surveyor Weekly*, 17 April 1985.
Barry, P. 'Getting involved: the QS in engineering'. *Chartered Quantity Surveyor*, July 1985.
Boyd, M. 'The QS role in rehab'. *Chartered Quantity Surveyor*, October 1982.
Brandon, P.S. (ed.) *Quantity Surveying Techniques: New Directions*. Blackwell Scientific Publications, 1992.
Dann, C. 'The responsibility of a profession'. *Chartered Quantity Surveyor*, December 1983.
Davies, C. 'Major consultants 3: The quantity surveyor'. *Architects Journal*, April 1984.
Davis, Langdon and Everest *QS 2000: The Future of the Chartered Quantity Surveyor*. The Royal Institution of Chartered Surveyors, 1991.
Feigil, B. 'Quantity surveyors in Europe'. *Chartered Quantity Surveyor*, November 1981.
Flanagan, R. and Norman, G. *Life-Cycle Costing for Construction*. RICS Books, 1983.
Gilbert, G. and Richardson, P. 'Investment appraisal'. *Chartered Quantity Surveyor*, August 1984.
Gow, H. 'Keeping up maintenance'. *Chartered Quantity Surveyor*, November 1985.
Hay, H. 'Making the best use of resources: the QS role in successful engineering contracting'. *Chartered Quantity Surveyor*, October 1982.
ICE *Civil Engineering Procedure*. Institution of Civil Engineers, 1992.
Knowles, R. 'Arbitration – where does the RICS stand?' *Chartered Quantity Surveyor*, February 1984.
Law, A. 'Life-cycle costing in the USA'. *Chartered Quantity Surveyor*, April 1984.
Lewis, B.J. 'The quantity surveyor in a new suit'. *Consulting Engineer*, October 1984.
Livingstone, W. and Easton, C. 'Audit and the quantity surveyor'. *Chartered Quantity Surveyor*, May 1983.
Marshall, R. 'Managing seabed operations'. *Chartered Quantity Surveyor*, August 1983.
Meopham, B. 'The QS in the management team'. *Chartered Quantity Surveyor*, September 1982.
Meopham, B. 'Cost control for the civil engineering contractor'. *Chartered Quantity Surveyor*, June 1983.
RICS *A Study of Quantity Surveying Practice and Client Demand*. The Royal Institution of Chartered Surveyors, 1984.
RICS *Client Guide to the Appointment of a Quantity Surveyor*. RICS Books, 1992.
RICS *Quality Assurance: Guidelines for the Interpretation of BS 5750 for Use by Quantity Surveying Practices and Certification Bodies*. The Royal Institution of Chartered Surveyors, 1990.
RICS *Review of Educational Policy*. The Royal Institution of Chartered Surveyors, 1978.

RICS *Surveying in the Eighties*. The Royal Institution of Chartered Surveyors, 1980.

RICS *The Future Role of the Chartered Quantity Surveyor*. The Royal Institution of Chartered Surveyors, 1983.

RICS *The Future Role of the Quantity Surveyor*. The Royal Institution of Chartered Surveyors, 1971.

RICS *UK and US Construction Industries: A Comparison of Design and Construction Procedures*. The Royal Institution of Chartered Surveyors, 1979.

RICS *What Does a Quantity Surveyor Do?* The Royal Institution of Chartered Surveyors, 1986.

Smith, J. 'Quantity surveyor's emerging role during the briefing stage'. *Building Economist*, March 1982.

Smith, L. 'Monitoring American construction costs'. *Chartered Quantity Surveyor*, April 1982.

Soloman, G.S. 'Cost management by quantity surveyors in the chemical industry'. *Cost Engineer*, Vol. 22, No. 5, 1984.

Spedding, A. 'Management of building assets'. *Chartered Quantity Surveyor*, June 1984.

Spencer, D. 'Cost control in engineering'. *Chartered Quantity Surveyor*, September 1984.

The Core Skills and Knowledge Base of the Quantity Surveyor. RICS Research Paper No. 19, The Royal Institution of Chartered Surveyors, 1992.

Welford, S. 'Total cost management'. *Chartered Quantity Surveyor*, July 1983.

Wheatley, G.F. 'The chartered surveyor in Europe: quantity surveying'. *Proceedings of the Annual Conference of The Royal Institution of Chartered Surveyors*, Jersey, 1979.

Williams, A. 'Quantity surveying in the US'. *Building*, March 1983.

Willis, C.J. and Willis, J.A. *Specification Writing for Architects and Surveyors*. Blackwell Scientific Publications, 1991.

Willis, C.J. 'Elements of quantity surveying in the future'. *Chartered Quantity Surveyor*, January 1979.

Wilson, A. 'Where we stand: the role of the quantity surveyor'. *Chartered Quantity Surveyor*, December 1984.

Wilson, G. 'Getting value for money overseas'. *Chartered Quantity Surveyor*, January 1984.

Yates, A. and Gilbert, B. *The Appraisal of Capital Investment in Property*. RICS Books, 1989.

Chapter 3

Private and public practice

Introduction

The vast majority of quantity surveyors are employed in private or public practice or in a contractor's organization. The last of these is the subject of Chapter 4. In addition, quantity surveyors have also been appointed in a variety of executive positions throughout the construction and other industries. In many instances, although their basic training as a quantity surveyor has been an asset in attaining a particular position, the role they now perform may well have little or nothing to do with surveying practice.

Traditionally, the principal difference between private and public practice was that whereas private practices are profit-motivated businesses, the main function of public practice was to ensure the accountability of public finances.

This difference has narrowed considerably over recent years with the privatization of many local and central government quantity surveying departments and former statutory authorities. The aims of the two sectors have therefore come together to a certain extent. They compete against each other for work, and their livelihoods also now depend upon profitability.

The status of the quantity surveyor is such that there is a need to provide clear, impartial and unequivocal professional cost and contractual advice. This means that a balance has to be achieved between maximizing profits and maintaining a duty to clients. For public officers this also includes accountability for public spending.

Private practice

Most quantity surveying firms have traditionally practised as partnerships. However, following changes in the RICS by-laws, some practices have elected to operate as limited-liability companies. The principal differences between the two are summarized in the following sections. The specific aspects of organization and management of the business are the subject of Chapter 5.

Partnership

Where two or more persons enter into partnership they are jointly and severally responsible for the acts of the partnership. Further, they are each liable to the full extent of their personal wealth for the debts of the business. There is no limit to their liability, as there is for directors of a limited company, to whom failure may only mean the loss of their shares in the company.

All partners are bound by the individual acts carried out by each partner in the course of business. They are not, however, bound in respect of private transactions of individual partners.

Partnerships come into being and expand for a variety of reasons.

- As a business expands there is a need to divide the responsibility for management and the securing of work.
- Through pooling of resources and accommodation economy in expenditure can be achieved.
- The introduction of work or the need to raise additional capital may result in new partners being introduced.

The detailed consideration of partnerships falls outside the scope of this book. Suffice it to say that while it is not a legal necessity, a formal partnership agreement should be in place, which legally sets down how the partnership will operate and covers such details as partners' capital and profit share.

Limited liability

Formerly the by-laws of the RICS prohibited members from parting with equity or shares to parties not actively involved in the practice. Dispensation to practise with limited liability could be granted subject to certain capitalization requirements.

Increasingly, however, it was felt that chartered quantity surveyors should be able to structure their practices so as to allow them to raise finance in ways that would enable them to improve their efficiency and effectiveness. It was also considered that both large and small firms would benefit from such a change and be better able to compete more effectively with other organizations, which represent a fast-growing competition in their markets. The view expressed by many was that to force chartered quantity surveyors to practise in partnerships was an unnecessary restriction that was unsupportable in this day and age. Consequently, in 1986 the RICS removed the restrictions on limited liability. The changes involved the following:

- removing the requirement for issued and paid-up share capital of the company to be not less than £25 000;
- Bringing the professional indemnity insurance requirements for surveyors practising with limited liability into line with the requirements of other surveyor principals;
- removing the restrictions on the transfer of outside share capital in surveyors' limited-liability companies, and on accepting instructions from outside shareholders;
- removing the requirement for surveyors wishing to practise with limited liability to apply to the Institution for permission to do so;
- imposing on surveyors who are directors of either limited or unlimited

companies a requirement to ensure that they have full responsibility for
professional matters;
- Imposing an obligation on such surveyor directors of companies to include
  clauses in their memorandum and articles of association which provide for
  matters to be conducted in accordance with the Institution's Rules of
  Conduct.

Several leading quantity surveying practices had already begun to practise as
limited-liability companies prior to these changes being implemented. Dereg-
ulation allowed them a great deal more flexibility in the running of their
companies, and led to more practices electing to operate in this way.

Public service practice

The same general principles of practice and procedure apply to both private and
public practice, with the obvious exception of the financial responsibility of the
principal and the differences in the character and requirements of the respective
clients.

For many years the amount of building work for which public authorities were
responsible grew continually, until by the mid-1970s it covered nearly 50% of the
nation's construction output. Since then the workload in this sector has fallen off
considerably (see Chapter 1). This fall is due, for example, to the reduction in the
requirement for public buildings such as government offices, police stations, fire
service buildings and schools. In addition, public housing is now to a large extent
handled by housing associations. A more recent factor has been the privatization
of local and central government quantity surveying departments and other
bodies, as referred to later. The effect of all this has been a large reduction in the
number of persons employed in public service. However, the service still exists,
and what follows highlights the significant differences between public and
private practice.

Most quantity surveyors, at some stage in their careers, are likely to find
themselves working for a public authority, whether in private practice or in the
public service. Because public authorities are spending public money, they have
elaborate (and often cumbersome) methods of administrative and financial
control and it is these, together with the size of the authorities, which affect the
surveyor's practice. Individual authorities and their quantity surveying
departments vary greatly, and it is only possible to provide a general description.

Scope of public practice

The public service includes employment in government departments, government agencies, local authorities and a number of statutory bodies.

Government departments and agencies include:

- the Department of the Environment;
- the Department of Transport, which controls road building;
- the Department for Education, which controls building of schools by local
  authorities and universities;

- the Department of Health, which controls the building activities of local health authorities (although their influence is decreasing as more self-governing Trusts are established);
- Property Holdings, which is responsible for the government's common user estate;
- Defence Works Services, which is responsible for Ministry of Defence building;
- the Lord Chancellor's Department, which is responsible for court buildings.

The volume of work constructed and controlled by government departments and particularly the government agencies involves a limited number of in-house professional staff including quantity surveyors, who deal with the control of projects varying in size from the very large to small minor works, as well as extensive maintenance and repair programmes.

Local authorities' offices vary considerably in size according to the areas governed. The government of London, the largest and the most complex urban area, is administered by the City of London and the London boroughs. Following the disbanding of the former Greater London Council, its responsibilities for building control, fire and ambulance stations, refuse disposal and roads in the Greater London area have passed to the boroughs or have been taken over by special bodies set up to deal with these matters. The majority of the boroughs have architectural departments to undertake the building responsibilities arising from the various services that they provide, although as described later this is now changing.

This is also the case with the other large city councils as well as the counties and districts, whose building requirements are mainly concerned with such things as local government offices, housing, schools and old people's homes.

Statutory bodies include national undertakings such as British Coal, British Rail, health authorities and the Post Office, who act as government-sponsored corporations. The nature of the work undertaken is naturally confined to the requirements of the particular organization, and in many cases civil engineering forms a large part of their work. Some former statutory bodies, such as British Telecom and the British Airports Authority, have now been privatized and more are likely to follow.

Organization

Public practice tends to reflect the organization of the large private firm, and staff are subdivided by function; again, the general organization and management aspects are covered in Chapter 5.

The difficulty of private practitioners spreading their work evenly is solved in the public service by engaging only a proportion of the staff required to deal with the total requirements of the particular office. The remainder of the work is let to private firms.

The range of quantity surveying functions undertaken by public practice offices varies according to policy, volume of work and staff available. The function most commonly assigned to the private practitioner is the preparation of contract documentation, but any or all of the functions may be so assigned whether for a proportion or for all projects. In some instances the estimating, cost planning, valuations and final accounts are carried out by the department surveyors, but only a small proportion of the contract documentation required is prepared by its own staff. At the other end of the scale the surveyors in some offices act as coordinators of the activities of private surveyors, to whom all work is let.

The introduction of fee competition and the requirement to comply with the EC Directives on Public Supplies, Works and Services together with the government's compulsory competitive tendering procedures will ultimately lead to all work being advertised and fee bids invited. This will also extend to all public departments themselves having to tender for work.

Conditions of employment

In private offices details of a surveyor's remuneration are a matter arranged between surveyor and employer. In public offices a surveyor's salary is dependent on the grade of appointment, and details of the salary scales are public knowledge. Each grade has a minimum and maximum and in many authorities there is provision for frequent reassessment of salaries to take account of changes in responsibility, additional experience and qualifications. Alterations to or improvements of the pay and grading structure are the subject of negotiations between representatives of a staff organization and the employer's or 'official' side. In most local authorities the negotiating body is the National Association of Local Government Officers (NALGO) and in the Civil Service it is the Institution of Professional Managers and Specialists. These bodies look after the interests of professions other than surveyors. Most appointments and promotions in the public service necessitate formal interviews before an examining board.

Duties

The work carried out by surveyors in public practice is technically the same as that of surveyors in private practice. They prepare approximate estimates, cost plans, documentation, and valuations for certificates; they measure variations and agree accounts with the contractor.

In 'controlling' government departments the surveyor's work is restricted to the examination of estimates prepared by surveyors for subordinate authorities, and of the tenders subsequently received with a view to the department's approval being given.

In recent years, surveyors in both 'controlling' and building departments have undertaken an additional and highly important task: the setting of cost limits arrived at in collaboration with administrative and architectural colleagues and other professions.

Forms of contract

Most local authorities and statutory bodies use the Standard Form of Contract or one based on that form. Central government work, however, is carried out under the GC/Works/1 Form, which may be used for both building and civil engineering work. Under this form of contract, the quantity surveyor has duties similar to those under the Standard Form.

No matter how elaborate or comprehensive a building contract is thought to be, there are frequently unforeseen circumstances, which require individual judgement and the client's approval. When dealing with an individual private client it is easier for the surveyor to discuss the matter with a view to recommending a course of action. The client is not likely to be creating a precedent and will therefore generally make a quick and usually reasonable decision. With government and, to a lesser extent, local authority work, the quantity surveyor's recommendation (whether it be a private firm or staff surveyor) has to be referred to a contracts directorate or a committee. Because they are spending public money and are subject to the scrutiny of independent auditors, and because they may be called upon publicly to account for apparently unusual expenditure, quantity surveyors must expect others to examine their recommendations in considerable detail. Whereas the individual client may be ready to meet a payment that is justifiable on moral grounds, officers in public service who exercise the client's functions will be reluctant to depart from the strict interpretation of the contract, irrespective of the surveyor's recommendation. When they do apply these judgements, the payment is invariably described as *ex gratia* to avoid creating a precedent.

Public service as a client

Such differences as exist between practice for a private client and for a public body largely result from the safeguards that are required in spending public money. Because of this, administrative controls are set up to consider, examine and approve building projects and their estimated costs. This creates a requirement for professional people, including surveyors, to submit proposals in a prescribed manner. Sometimes they also have to perform duties that would not be required by private clients.

When it comes to accepting tenders, honouring certificates and paying final accounts, methods of financial control are required to make sure that every payment is properly authorized and within the amount approved, and that there is no possibility of fraud or wilful negligence.

The administrative and financial controls are operated by administrative staff and accountants who have no training and little knowledge of quantity surveying or building contracts, beyond what they gradually acquire through experience. Yet they find themselves answerable to committees and district auditors in local government or, in central government, to the Treasury, Select Committee on Estimates, Exchequer and Audit, the Public Accounts Committee and even Parliament itself. It is often the case that what seems a straightforward case to the surveyor takes a long time to reach settlement. The system of controlling public expenditure, particularly in government service, is over 100 years old, and some administrators and economists think that fundamental changes are now due. Whatever the procedure, the surveyor must provide the information, advice and help considered necessary in the interests of the public purse.

Two requirements of the public service as a client particularly affect the surveyor, and no distinction is made between the surveyor in private practice or in public service.

When a project is financed from public funds a specified sum is allocated for the purpose. The sum is set within cost limits, which should not be exceeded, and the surveyor is expected to assist the architect or engineer in keeping the cost of the project within the allotted sum. If additional money is required, a case has to be fully substantiated and approval obtained before the authority is committed to the additional expenditure. This is particularly important when successive Chancellors keep trying to reduce expenditure. With private clients, although the aspect of cost control is of equal importance, the obtaining of additional funding is unlikely to be subject to such rigorous procedures.

The second requirement that affects the surveyor closely is that of audit. Most public offices, other than government departments, have their accounts audited twice a year, once internally by members of the finance officer's staff and again externally by independent auditors. The officers who carry out the audit are mostly accountants, but others without financial or quantity surveying training are also employed. Their task is to ensure that the financial provisions of the contract have been faithfully carried out, that all payments have been properly made, and that any unexpended allowances have been recovered.

The purpose of auditing is to ensure that the correct procedures have been properly followed. In addition, the auditors must be aware of the possibilities of negligence or fraud. Because of these responsibilities it is necessary for quantity surveyors to prepare the final account in strict conformity with the conditions of the contract. They do not have the same latitude as with a private client, and cannot always use the 'give-and-take' methods of balancing trifling or obvious self-cancelling variations. Only an experienced quantity surveyor can judge the fairness of 'give-and-take' methods, and the auditor cannot be expected to have this skill. As public money is being spent, each account and payment must not only be correct but be seen to be correct.

Comparison of public and private practice

The surveyor in public service does not have the same anxieties as those in private practice, such as finding capital and offices, ensuring a flow of work and avoiding losses. Once public departments have been privatized, however, these factors need to be considered.

Though the variety of work in some public offices may be limited, there is some satisfaction in being concerned in a continuous programme of national or local public works and, though the 'client' may always be the same, at least the surveyor has a foreknowledge of requirements. In the larger offices the work is varied, but if the staff is specialized, surveyors may be confined to a limited range of duties, which may tend after a while to make them feel they need different experience. There is, of course, the same possibility in the larger private practices. In such circumstances surveyors gain a detailed expertise in terms of funding and economics and the financial consequences on particular types of construction projects.

The surveyor's great responsibility in the public service is that of controlling the expenditure of public money, money placed at the disposal of departments by people who, unlike private clients, are quite unknown and who have no option but to assume that their money is being spent wisely and economically.

Bibliography

Ashworth, A. 'The auditing of building contracts'. *QS Weekly*, 14 December 1979.

CIPFA *Audit of Building and Civil Engineering Contracts*. Chartered Institute of Public Finance and Accountancy, 1973.

Cripps, Y. 'The professions: a critical view'. *Arbitration*, November 1986.

Male, S. 'Professional authority, power and emerging forms of "profession" in quantity surveying'. *Construction Management and Economics*, Spring 1990.

Palmer, J. 'Professionalism, standards and negligence'. *RICS Annual Conference*, 1982.

RICS *A Study of Quantity Surveying Practice and Client Demand*. The Royal Institution of Chartered Surveyors, 1984.

Young, A. *The Manager's Handbook*. Sphere Books, 1986.

Chapter 4

Contracting surveying

Introduction

The organization of building and civil engineering companies varies considerably from firm to firm. Some of the larger firms, for example, may be truly general contractors, while other firms with comparable turnovers of work may do very little of the construction work themselves but rely almost entirely upon subcontractors. The smaller firms will often expect a wider range of skills from almost everyone they employ. The contractor's surveyor in some companies may therefore undertake a rather narrow or specialized range of tasks, and in other firms may be expected to undertake work that is normally outside the periphery of quantity surveying. The size of the contractor is therefore a very important influence on the surveyor's work.

The second important factor affecting the surveyor's work is the management structure of the firm. In some companies the whole of the quantity surveying is seen as a separate function and is under the direction of a surveying manager. In other firms the quantity surveyors work with other disciplines under the authority of a contract manager.

Conditions of employment

The conditions of employment for the majority of quantity surveyors working for contractors are somewhat different from those of their counterparts employed in professional practice or public service. Perhaps the most obvious difference is that contractor's surveyors spend more time on the construction site. In many cases they may be resident surveyors on a single site, or use this as an office for looking after a number of smaller contracts. Their working environment is therefore not as comfortable as it is only of a temporary nature, and the surroundings are more akin to industrial conditions.

Being a member of the contractor's project team also means more interdisciplinary working at all levels of the surveyor's career. They may for example share an office with someone from a different profession, and they therefore have the opportunity of gaining a better understanding of other people's work.

Many contractors also expect their quantity surveyor to work the same hours as their other employees on site. This often means an earlier start in the morning and a longer working week than those surveyors employed by the professional consultants. Furthermore, it is unusual to receive any overtime payment on those occasions where it is necessary to work late in the evenings or at weekends.

Contractors' surveyors do, however, tend to enjoy greater freedom to do their work, and may also receive extra responsibilities earlier in their careers. There is also a better chance of being given a company car, particularly where the surveyor has to look after a number of sites. Some contractors will in addition pay a site allowance, especially in those circumstances when the surveyor has to live away from home. Contractors' surveyors are generally thought to be financially better off than surveyors working in private practice. This will, however, vary at different times during a career, and will need to take into account the various differences in the conditions of employment.

For a contractor's surveyor with one of the large national companies, there may also be the disadvantage or advantage (depending on how one looks at it) of being more mobile. It will be necessary to move to where the work is, and this could require the family moving house frequently throughout a career. The building industry has now become much more regional in nature and it may therefore be reasonable to expect only to work on sites within a certain location. Civil engineering is more nationally organized, but the larger projects may tend to offset the otherwise more frequent mobility. Contractors' surveyors working on industrial engineering projects are often confined to those areas where these industries are located.

Role

The role of contractors' quantity surveyors is somewhat different from that of the professional or client's quantity surveyor. They are unashamedly more com-mercially minded, and sometimes the financial success or failure of a project or even a company is due in part at least to the contractors' surveyor. While the client's quantity surveyor may claim impartiality between the client and the contractor, the contractor's surveyor will be representing their employer's interests. Prudent contractors have always employed quantity surveyors to look after their interests, and have particularly relied upon them in the more controversial contractual areas.

Function

Contractors employ quantity surveyors to ensure that they receive the correct payment at the appropriate time for the work done on site. In practice their work may embrace estimating and the negotiation of new contracts, site measurement, subcontractor arrangement and accounts, profitability and forecasting, con-tractual disputes and claims, cost and bonus assessment, site costing and other matters of a management and administrative nature.

The following are some of the more common contracting surveying functions.

Estimating

In the larger firms the surveying and estimating departments are quite separate functions, each with its own manager or director responsible for its operations. It is now quite common practice in such firms for the estimating function to be computerized. This can then be linked to provide a valuation package, which in turn can be linked to cost systems of varying levels of sophistication. In the smaller firms these two departments may be one, with the surveyor then having the dual role and being responsible for estimating. Some surveyors will do rather more estimating than surveying and this key function is the start of the project's successful conclusion. Mistakes made at this stage are difficult to rectify later, and great care needs to be taken, particularly with pricing the major cost items of work. Successful estimating relies upon a good knowledge of the procedures used, the contract documentation, how and when the work will be carried out, the level of pricing in the market and the commercial state of one's competitors. The importance of estimating in practice is not so much to do with the mechanics of price analysis, as in winning the project and being able to carry out the work for the projected profit.

Negotiation

Contractors' surveyors need to be good negotiators. They will frequently have to agree prices with the client's quantity surveyor, sometimes before the contract is signed (in the case of a negotiated contract) and may be responsible for the preparation of final accounts or possibly contract claims. They will need to agree a fair sum to be included in a valuation, always trying to ensure that they get a fair recompense for their employers. The art of negotiation requires firmness, fairness, and knowing what one is trying to achieve. Negotiation always relies upon the assumption that there is a genuine desire for an equitable agreement.

Site measurement

Contractors' surveyors will spend a proportion of their time on site, measuring the work for different purposes. This may be done for such things as internal or external valuations, final accounts, subcontract payments, cost and bonusing. They therefore become more familiar with the project and conversant with the problems that occur. It also increases their proficiency in the skills of site measuring and the various instruments used for that purpose.

Financial management

A part of the contractor's surveyors' role is the financial management of the project. It is becoming more and more their responsibility to assess the financial performance of each contract, and to discover the loss-making operations. Where these can be found out early enough, there is the possibility of reducing the loss. Where such loss is due to the effects of either consultant or client intervention, then this may form the basis of a contractual claim. Financial management involves cost and value reconciliations over the whole of the contractor's operations and should be done at regular intervals. This information can then be reported back to the commercial managers for their information and decisions. These reports may indicate inefficient methods of working, poor estimating or inappropriate buying policies. They may also show that the contractor is more competent and profitable with certain types of job.

The assessment of the financial performance of a contract is simplified when a contractor makes use of an integrated estimating/valuation/costing computer system as referred to above. This should not only be much quicker than a manual system but should also immediately identify a problem trade or trades and then the individual component: labour, plant or material cost. This is a vitally important part of a contractor's surveyor's role. The information produced is, however, totally dependent on the accuracy of the data that is input. It is essential that labour, plant, material, subcontract and overhead costs are correctly allocated, and ideally the allocation or coding procedure should be initiated by the surveyor.

Financial management must not be looked upon solely as an historic activity, since it must be projecting forward targets by which to measure performance. The usual forecasting that management will require involves the preparation of monthly turnover and profit forecasts, and these must be carefully adjusted to take into account anticipated future costs and revenue. Care must be taken to ensure that the information provided is as clear as possible so that management can then take the appropriate action.

Interim payments

A principal duty of the contractor's surveyor is to prepare the application for interim payment. The private quantity surveyor sometimes prefers the valuation to be done in this way, since it saves a considerable amount of time when visiting the site. The contractor's surveyor must make sure that the amount is realistic and that it covers all the work executed during that month. Interim payments are the lifeblood of the contractor and any shortfall could have severe consequences on cash flow. The contractor's surveyor must therefore make sure that all relevant subcontractors' work has been included, and variations at least approximately remeasured; increased costs and contract claims, where partially agreed, are also part of the valuation. The value of materials on site can form an important part of the total valuation and wherever possible, and the contract allows, so can materials off site.

While the introduction of predetermined valuation payments based on charts or other means simplifies the computation of interim valuations, the contractor's surveyor still needs to know for internal costing purposes the value of work actually carried out, and in consequence still needs reasonably detailed measured information.

Subcontractors

An important duty for contractors' surveyors is dealing with the various types of subcontractor. They may in the first instance become involved in placing the orders with the subcontracting firms. With the directly employed subcontractor this involvement may commence with the initial enquiry made by the estimator. Once on site the surveyor will then be responsible for measuring, valuing and agreeing the subcontractor's work in accordance with the terms of the subcontract. Where variations arise the surveyor will also have the task of agreeing this work with the professional quantity surveyor. The financial control of subcontractors has become an increasingly important area of work for the contractor's surveyor, particularly as more of the work completed on site is now done in this manner.

Nominated subcontractors provide different sorts of problems for the contractor, as their selection is the responsibility of the contract administrator. The contractor is, however, responsible for managing them and ensuring that they comply with the conditions of the contract. Surveyors must try to make sure that their employer's rights are properly protected when accepting nominations.

Site costing

Costing is a control procedure used by the contractor's surveyor. It involves a comparison of what was planned with what has actually taken place. One of its main purposes is to reveal the efficiency or otherwise of specific activities, usually by comparison, and then to take any appropriate action that may be necessary. Any cost control system will have three facets;

- the selection of desirable and achievable targets for the work to be carried out;
- the comparison of actual performances with those targets;
- a means of corrective action to be taken should the performance fail to meet the designated targets.

Perhaps one of the biggest difficulties to overcome is that of the time-lag delay in the preparation and comparison of the performance data. A good cost-control system will therefore incorporate an early-warning system. In practice, emphasis is placed upon those site operations that involve the contractor in the largest expenditure. The site costing system will therefore provide the following:

- cost data for future estimating;
- budgets and targets for future work;
- highlighting of uneconomic methods of working;
- identification of excessive wastage or inefficiency in the use of resources;
- information that will enable the firm to undertake work that it is best equipped to carry out.

It is essential that all cost information flowing from the production of a valuation is available within one week; otherwise, corrective action to alleviate losses may not be possible before the completion of a loss-making operation.

Final account

Another important part of the contractor's surveyor's work is the agreement of the final account. Under the terms of the contract the private quantity surveyor is responsible for its preparation, but in reality the best approach is for both surveyors to work together to produce an agreed account. The contractor's surveyor should be alert at identifying variations and work that should be reimbursed on daywork rates.

The process of measurement should present only a few disagreements, perhaps over the interpretation of the rules of measurement. The agreement of rates for work not included in the contract documents may present more of a problem. The contractor's pricing notes at the time of tender will be helpful here, if they can be made available.

An inordinate amount of time is now spent in dealing with those matters where contractors feels that they have not been properly reimbursed under the terms of the contract. This can be a very time-consuming part of the contractor's surveyor's work, to such an extent that some surveyors are solely retained for this purpose. The contractor's surveyor should strive as far as possible to alleviate any possible delays in agreeing the final account.

The introduction of predetermined costing of variations – the provision of a firm estimate before the work is ordered – requires the same attributes but applied somewhat earlier in the process. Contract conditions require the anticipated cost of any disruption or prolongation to be taken into account when precosting variations and this involves imaginative anticipation of what may or may not happen in the future (see Chapter 11).

Contractual matters

Contractor's quantity surveyors are generally regarded as advisors or experts on contractual procedures and interpretation within their organization. They must therefore have a good working knowledge and understanding of legal processes and the framework of law, the standard forms of contract and subcontract that are generally in use, and relevant legal case law. They must ensure that they keep up to date with the various changes that frequently occur, and are aware of future developments and their effects upon the contractor. They must also be able to express their opinion clearly and effectively when discussing matters of a contractual nature with consultants or subcontractors.

Design and build

There are now many construction companies who are actively engaged in design and build, speculative development and similar sorts of ventures. Contractor's surveyors in these companies are likely to become more involved in precontract work in some way or other. They may be responsible for advising the designer on the cost implications not only of the design but also of the methods which the firm may use for construction on site. They will also be consulted on the contractual methods to be used and the documentation that is required, and will recommend the ways in which the scheme might be financed, taking into account the speed of completion.

Work for subcontractors

Since a large amount of work is now undertaken by subcontractors, this section of the construction industry can benefit from using quantity surveyors. The surveyors who work for these companies will undertake a far wider range of duties and may well be concerned with all aspects of a financial nature, including negotiating with the bank, the submission and agreement of insurance claims, VAT and dealing with the companies' accountants. In other capacities they may be responsible for general management, the allocation of men to projects, planning their work and dealing with their wages.

Education and the contractor's surveyor

Within the UK, and in several other countries, courses are offered at various levels leading to qualifications in quantity surveying. These include degree and professional institution courses. Management studies have been incorporated into the curriculum of these courses but in some cases have been restricted solely to business management, while others have seen fit to incorporate subjects such as contract planning and control.

Quantity surveyors employed by contractors may decide to become members of The Royal Institution of Chartered Surveyors or other institutions such as the Chartered Institute of Building or the Association of Building Engineers. These institutions claim a better representation and recognition within them for contractors' surveyors and seek to emphasize the different role which they have to perform. The RICS has, however, fully recognized the involvement of quantity surveyors in contracting companies and such surveyors play an equal part, with their professional colleagues, in the running of the Institution.

As far as the contractors are concerned, they favour staff who can undertake a wide range of surveying work completely. They do, however, also prefer staff to have professional qualifications, realizing that the knowledge and expertise gained through study is of direct benefit to their work and to the profitability of the company.

Relationships and the client's quantity surveyor

The contractor's surveyor's first contact with the client's quantity surveyor will probably be in raising queries on the contract documentation, particularly if it is the surveyor who has priced the work. Upon receipt of the acceptance of the contractor's tender, the contractor will inform the architect of the names of the agents, contracts manager, quantity surveyor and other senior personnel who will be responsible for the project.

The contractor's surveyor and the quantity surveyor would then normally meet together to discuss points that were raised at tender stage or during a post-tender meeting. These would include matters relating to the correction of errors in pricing, the procedures to be followed regarding remeasurement and financial control, and the dates of interim valuations.

The relationships between the two surveyors will usually start amicably enough, and this is the way it usually ends, but there is always the possibility of deterioration in the relationship developing. This may occur because one party does not feel that the other is being as fair or as scrupulous as they ought to be.

It needs to be remembered that the two surveyors are trading in a very desirable commodity: money. While fairness is an excellent motive to follow and should help to reduce the number of potential problems arising, it must also be remembered that there is a contract to be considered, and auditors will need to be satisfied that correct procedures were followed. Each party to the contract – the client and the contractor – have employed their respective surveyors to ensure that their own interests are properly protected. On matters of dispute the advice may well be to settle these differences on a give-and-take basis.

It is all too common nowadays for disputes to lead to arbitration or indeed to litigation. Before this happens both surveyors should warn their employers of the considerable financial costs of legal disputes.

Future prospects

The future prospects for the contractor's surveyors are seen to be very good, for several reasons. First, the increase in the direct employment of contractors for design-and-build or management contracts places a greater onus and more emphasis on their work. In these circumstances it may become necessary for them to become more involved in precontract activities. Secondly, because contracting margins are very tight, companies tend to rely upon the surveyor's expertise to turn meagre profits into more acceptable sums, and loss-making contracts into profit. Thirdly, because of their proven performance there are now wider opportunities for them amongst a whole new range of firms who require their skills.

There is also new evidence of contractors' surveyors being employed at the highest levels within firms, which in itself emphasizes their importance and prospects for the future. The majority of the larger contracting firms will now have at least one director whose professional origins have been in quantity surveying. Such a director may be responsible for the surveying and estimating departments in the company or perhaps more commonly today will have a title such as Commercial Director and is likely to be concerned with the financial decisions and management of the construction company. Many medium-sized firms have seen the importance of giving their chief quantity surveyors considerable responsibility in financial matters, with their importance overall in the firm being recognized by a directorship.

Bibliography

Barrett, F.R. *Cost Value Reconciliation*. Chartered Institute of Building, 1981.
Buchan, C., Fleming, M. and Kelly, J. *Estimating for Builders and Quantity Surveyors*. Butterworth, 1991.
CIOB *Code of Estimating Practice*. Chartered Institute of Building, 1983.
Cooke, B. and Jepson, W. *Cost and Financial Control in Construction*. Macmillan, 1982.
Cottrell, G.P. *The Builder's Quantity Surveyor*. CIOB Surveying Information Service No. 1, Chartered Institute of Building, 1980.
Cross, R. *Builder's Estimating Data*. Heinemann, 1990.
Doland, D. *The British Construction Industry*. Macmillan, 1979.
Goodlad, J.B. *Accounting for Construction Management*. Heinemann, 1974.
Johnston, N.G. 'Directors in construction companies: another role for the quantity surveyor'. *Proceedings RICS Triennial Conference*, 1978.
Kelsall, J.J. 'A contractor costing system'. *The Project Manager*, January 1978.
Nedved, S.C. *Builder's Accounting*. Butterworth, 1983.

Norman, A. 'Cost control and cash flow accounting'. *Project Manager*, January 1983.

Pearson, N. 'The control of subcontractors'. CIOB Occasional Paper No. 7, Chartered Institute of Building, 1975.

Skinner, D.W.H. 'The contractor's use of bills of quantities'. COIB Occasional paper, Chartered Institute of Building, 1981.

Trimble, E.G. and Kerr, D. 'How much profit from contracts goes to the bank?' *Construction News*, March 1974.

Chapter 5

Organization and management

Introduction

Setting up and expanding a practice or business, together with general management and finance, are all subjects adequately dealt with in the many textbooks on business management. However, there are certain specific aspects of organization and management relating to the work of the quantity surveyor that are worthy of consideration in some detail. They include:

- staffing
- office organization
- marketing
- management of quality, time and cost
- education and training
- finance and accounts.

Staffing

The staffing structure within a quantity surveying practice, public service or contracting organization will comprise members of staff carrying out the quantity surveying or other specialist service and those providing the necessary support services such as administration, accounts and information technology. The overall staffing structure will differ from one organization to another, but certain general principles will be similar.

Private practice

The way an office organizes and carries out its surveying work will to some extent be determined by the size of the practice and the nature of its commissions. The larger the practice the more specialized the duties of the individual surveyor are likely to be. There are essentially two modes of operation, as shown in Fig. 5.1. Type (a) separates the normal surveying work on a project into the cost planning, contract documentation, final accounting and specialist services. Some coordination will be necessary between these phases, as the work in each will be carried out by different personnel. This type of organization gives staff the opportunity to develop their expertise in depth.

Type (b) allows the surveyors to undertake all aspects of the project from inception to final completion and provides the surveyor with a clearer under-

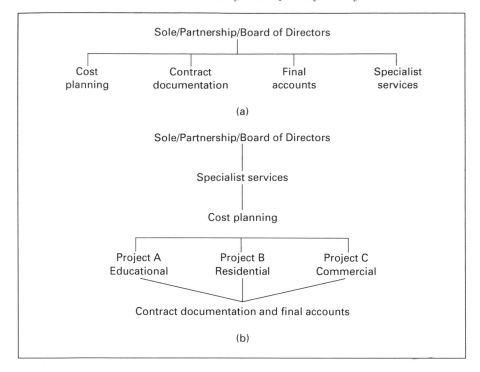

Fig. 5.1   Modes of operation of office organization.

standing of the project. This alternative will often allocate the staff to teams, who then in turn become specialists in certain types of project, depending of course on the type of work undertaken and the workload at any one time.

Most practices today also carry out other work that may be incidental to their main work, possibly because of the interest or particular skills and knowledge of members of staff. In other instances specialist work represents a major source of work for the practice. Such specializations are covered in Chapters 12–15.

Whichever procedure is used by the office for carrying out its work, some programme of staff time will be necessary if the services are to be carried out efficiently to meet the project requirements. The first management aid needed to plan and control the work is a planning and progress chart. This is a linear timetable, which breaks down the project into its various parts, and to be complete it must take into account all aspects of the project, including the preparation of interim valuations and the final account. It may be desirable in the first instance to prepare the chart on the basis of precontract work alone but for overall planning purposes some account must also be taken of the other duties that have to be performed. The type, size and complexity of the project will need to be taken into account in planning the work that has to be done.

The overall management of the practice and the well-being of the employees are the responsibility of the partners/directors. Each project will be the responsibility of an associate partner/director, or at least one of the senior surveying staff who, having identified the work that needs to be carried out within a particular time period, will determine the resource level required to perform the various tasks, either using staff within the office or calling on freelance staff as necessary. It is essential to preplan resources in order to budget costs against anticipated fee income. A major problem often facing the practice is the restricted time available to produce tender documentation and the like. Being at the end of the design process, they have to rely upon the other members of the design team to supply their information in good time. This emphasizes the importance that should be attached to an adequate programme of work, by identifying the work that has to be done, who will do it, and the date by which it should be completed.

Contracting organizations

The organization of building and civil engineering companies is briefly described in Chapter 4.

Support services

In addition to the surveying and other specialist staff within an organization there will also be the support staff necessary for its efficient operation. These will include secretarial and administrative support as well as the finance and accounts and information technology departments (information technology is covered in Chapter 7). However, the ever-increasing influence of computer systems on the quantity surveying and office administrative functions means that the balance of the general staffing structure is continually changing.

Office organization

General organization

Quantity surveyors, whether in private practice or in a support department within a contracting organization, provide a service, and therefore the direct staff costs and indirect costs (such as employers' National Insurance contributions and pensions contributions) form the largest part of the cost of running the practice or department.

The general establishment, expanding and equipping of an office are outside the scope of this book. However, in addition to staffing costs there are significant further costs associated with the running of an office, and it is worth noting the main items that constitute the general office overhead.

The major component of overhead expenditure is likely to be the office accommodation in terms of rent and the cost of furniture and equipment (including IT equipment) and communications necessary for the office to function efficiently. Other items of overhead expenditure are likely to include administrative costs, marketing, and general expenses.

Specialist stationery

In addition to general stationery, there is specialist stationery for use by the surveying staff in performing various tasks. This includes paper for abstracting and scheduling, approximate estimating, single and double cash column bill paper and dimension books. Traditional dimension and cut-and-shuffle paper, although still used, have in many cases been made redundant by the development of computer billing and word-processing systems.

Reference books and information

A characteristic of a quantity surveying office or department is the specific need for reference books and cost and technical information to assist the surveyors or specialists in performing their duties. Reference libraries will have been developed over a number of years and it is important that they be kept up to date to ensure that the staff can be fully aware of current developments in all aspects of the construction industry.

The core material that is required for reference, and which will need frequent updating, includes the following:

- textbooks on the principal subjects encountered in the office
- *British Standards Institution Handbook*
- construction price books
- BCIS cost information (also available as a computer application)
- current editions of all forms of contract likely to be adopted
- wage agreements
- plant hire rates
- EU directives/publications
- professional and general construction industry journals
- technical literature on products, materials and components
- specifications.

Building costs records

A certain amount of keeping of records is essential in any office, though there is a tendency to delay preparing them because they are not directly productive. They have a habit of never getting done when the information is fresh. The surveyor's own cost records will be of much greater value than any amount of published cost data. They relate to live projects and can be stored in a manner that easily facilitates retrieval. Some care, however, needs to be exercised when using them because there could have been special circumstances which may or may not need to be reflected in any re-use of the information. It is, however, more likely to reflect the pricing in the local area and therefore in this context it is more reliable.

It is good practice to prepare a cost analysis of every tender received, and these analyses will form a bank of useful cost records. If prepared in accordance with the Principles of Analysis prepared by the BCIS then some measure of comparability with their cost analyses can be achieved. Different clients may require different forms of cost analysis to suit their particular purpose. Supplementary information on market conditions and specification notes should also be included. These will be necessary in any future comparison of schemes.

It is important when compiling cost records to maintain a constant basis. The use of the BCIS principles will greatly assist in this. For instance, it may be desirable to omit external works, site clearance or preparatory alteration work from the price. The analysed tender sum must, however, make this omission clear.

Refurbishment projects do not lend themselves so well to records of this kind, and it is therefore preferable to provide an analysis in a rather different way, perhaps by identifying the cost centres for this type of project. Although these types of scheme are commonplace today in the UK, a standard form for their cost analysis has yet to be devised.

When the building scheme is complete a final cost can be recorded, and this may also be calculated as a cost per functional unit or superficial floor area. No attempt should be made to recalculate the cost analysis, as the project costs by now will be largely historic and the allocation of the final accounts to elements is very time-consuming.

BMI have a form of cost analysis for the maintenance and running costs of buildings. Surveyors concerned with costs-in-use or the occupancy costs of buildings may need to refer to or prepare analyses in this way.

Employer's responsibilities

There are certain health and safety and insurance issues to be addressed by employers.

Health and safety

There are three main statutes concerned with health and safety matters that are relevant to offices.

The Health and Safety at Work etc. Act 1974 imposes general responsibilities on the employer in respect of the safety, health and welfare of employees. The employer must ensure that any plant or equipment is properly maintained, that the systems of work are safe, and that entrances and exits are easily accessible. Provision must be made for instruction or supervision where this is required. The general public visiting the premises must also be properly safeguarded from any risks. Employers are under a responsibility to take reasonable care in these matters.

The Offices, Shops and Railway Premises Act 1963 applies to premises where an employee works for more than 21 hours per week. It includes the need to provide adequate heating, lighting, ventilation, sanitary conveniences, washing facilities and drinking water. The provisions lay down minimum space standards to prevent overcrowding, making allowances for space occupied by furniture, fittings and equipment. In some buildings some of the occupier's responsibilities are transferred to the owner of the premises, who is also responsible for complying with the provisions for the common areas such as entrances, passages, stairways and lifts. Environmental health officers from the local authority are responsible for enforcing the legislation in private offices. Offices in local authorities are the responsibility of inspectors from the Health and Safety Executive.

The Fire Precautions Act 1971 applies to offices and other designated premises where more than 20 people are employed. Note, however, that there is an obligation to provide reasonable means of escape in the event of fire for all office premises. Where premises fall within the Act an application must be made to the fire authority for a fire certificate. The authority will then visit the premises to assess the means of escape and other fire prevention measures and make recommendations where necessary.

Fire insurance
Insurance of the building will usually be the landlord's responsibility, but the tenant will need to cover the contents of the office against damage by fire and burglary. The general furniture and stationery will present no difficulty; however, a serious consideration in the event of a fire is the replacement of documents in the office, particularly those referring to contract documentation in the process of preparation. If everything is destroyed there may be no alternative but to begin again. The client will not of course pay a double fee. It is therefore advisable to insure the documents in the office under a special item. It will be very difficult to assess a figure, but it should be substantial, the premiums being a comparatively small matter when the contingent liability is considered. Only the cost of their replacement as it relates to salaries and overhead costs will be covered, unless a 'loss of profit' policy is also taken out.

Employer's liability insurance
An employer is liable to pay compensation for injury to any employees arising out of and in the course of employment, caused by the employer's negligence or that of another member of staff. In order to be able to meet these obligations the employer is required by law to take out a specific insurance. The Employer's Liability (Compulsory Insurance) Act 1969 and the Statutory Instrument (1971 No. 1117) making General Regulations require that every employer who carries on business in England and Wales shall maintain insurance under approved policies with authorized insurers. Premiums are based upon the type of employee and the salaries that are paid.

Public liability insurance
An owner or lessee of premises may be liable for personal injury or damage to the property of third parties caused by their negligence or that of a member of staff. The insurance cover should be sufficient to cover the different status of individuals, and actions by both principals and employees not only on the premises but anywhere while on business, and in countries abroad if this is appropriate.

Professional indemnity insurance
A conventional public liability policy does not normally cover any liability where professional negligence is involved. Reference is made in Chapter 6 to the RICS requirements regarding maintenance of professional indemnity insurance.

Marketing

In the current climate of recession within the construction industry the marketing of quantity surveying and other construction-related professions has taken on added importance. The marketing of professional services is a difficult task, however, as it involves selling a service rather than a product.

As quantity surveyors look to diversify the services they offer, it is necessary for them to recognize their specific strengths and weaknesses and those of their competitors, and to identify potential clients (in both existing and new market sectors) together with their needs and requirements. The services provided can then be tailored to meet these requirements and a marketing strategy can be targeted accordingly.

It is essential to have a clear marketing plan, to develop the marketing and presentational skills of key individuals within the organization, and to maintain an awareness of potential business growth.

Corporate image

Quantity surveying practices, whether consciously or not, present an image of themselves: both to others employed in the construction industry and to new and existing clients. This image may be developed from the way the firm's partners or directors see the future role of the profession of quantity surveying. Some practices are therefore content to see themselves in a traditional role, offering a service that is still demanded by many clients. Others have attempted to respond to the apparent changing needs of clients by demonstrating new skills and services and offering a much wider portfolio of advice and surveying services (see Chapters 12–15).

The image of the firm is therefore reflected to a large extent by the aspirations of the partners or directors and their collective experience, together with the commissions that they have undertaken. The expertise of the other members of the practice must also not be underestimated, as they are able to contribute to and enforce this image. It is important to have available details of staff development and achievements to reinforce the image. The important qualities to be portrayed are those of integrity, reliability and consistency.

The types of project, services provided and methods of working are important aspects to consider, and ones in which future clients are likely to be interested. Clients will also want to know how the practice keeps to programmes and costs and how good generally its advice has been. They will wish to identify its various strengths and weaknesses, and will be interested to know how the practice has managed changes in technology and perhaps how this will affect the future method of working.

The firm's track record, future potential and ability to solve problems should be conveyed in such a way that the practice demonstrates that it is the most appropriate to undertake the client's work, will suit the client's method of working, and will provide the quality of service commensurate with the fee charged.

Public relations and marketing

The surveying profession generally has done little in the past to advertise its skills to the public. The majority of the public therefore still presume that surveying equates with land surveying, and among the other established professions outside the construction industry quantity surveying is still largely unknown. The RICS Quantity Surveyors' Division in recent years has attempted to redress this imbalance by carrying out studies amongst clients and by publicizing its ever-widening range of services. The RICS on a much wider scale seeks to promote all the divisions, and has at times a difficult task in presenting a corporate image for the profession generally.

In 1991 the RICS published *Quantity Surveying 2000: The Future of the Chartered Quantity Surveyor,* which through research studies reviewed the changing markets, the changing nature of the construction industry, the changing profession and concluded with how the profession should respond to these changes through diversification into new markets and provision of new services.

The Institution has therefore now adopted a rather more positive and proactive role for its members than its hitherto passive position. It now seeks to ensure that the major financial institutions, development banks and aid agencies are continually aware of the surveyor's essential contribution to the development and construction process. It would be of even greater advantage to ensure that the quantity surveyor's services are properly understood and appreciated by those with whom contacts are first made. Client organizations must be encouraged to make full use of quantity surveying services and be convinced that they do provide value for money in all commercial and contractual aspects of construction.

The regulations on advertising and publicity for the function of quantity surveying have been significantly relaxed by the RICS over the years. However, care must still be taken to ensure that all statements made about the firm are an accurate representation. Advertisements must not seek to explicitly solicit instructions, nor compare a firm's services with those of others.

Practice brochures

The general easing of advertising restrictions presented both the quantity surveying profession and individual firms with new ways of marketing their services and seeking new clients. One of the most effective ways of achieving this is the preparation of a brochure that describes the practice and the services that it has on offer. It can be sent out in response to enquiries or distributed when the opportunity arises. It may be necessary in the first instance to prepare the brochure in conjunction with a marketing agency in order that the correct balance and effect is achieved.

The type of brochure available varies from something prepared on the practice's own word processor and assembled into an appropriate folder, to the elaborate, professionally prepared booklet, which comes complete with colour illustrations. Some practices provide a variety of literature and other material which seeks to emphasize the different specialist services that they offer. Other practices, in addition, provide a quarterly contract or cost review, which aims to keep their name firmly in their client's mind.

The production of a brochure must be done with care and skill to create the correct corporate view of the practice. It should not be dull and uninteresting but should give the appearance of something that demands to be read. The information must be factual, but its presentation should be persuasive and emphasize the qualities and skills of the practice. The information contained may include: the origins and development of the practice, partnership details, services provided, projects completed, office addresses, lists of clients and photographs of partners and projects.

Presentations

A considerable proportion of new commissions are obtained through fee competition, which often involves preparation of prequalification documents, details in support of a fee tender submission and in certain instances formal presentation to clients in person.

Although the format of presentations will differ depending upon the specific client, the project and the services to be provided, the main aim is to get across to the client general details of the practice, specific previous relevant project experience and (most importantly) the capability to carry out the required role.

Contractors are increasingly being required to prequalify for inclusion on tender lists. In addition to general company information and experience they need to identify their approach to the important aspects of management strategy and programming capability.

Quality management

Quality management systems

Considerable attention is given nowadays to the aspects of quality of the service provided by consultants and contractors to their clients. Increasingly clients in all sectors of the construction industry are demanding that consultants and contractors operate a quality management system and obtain third-party certification to demonstrate compliance with the British Standard for Quality Systems, BS 5750 (Part 1) (ISO 9001/EN 29001).

Practices and companies wishing to obtain certification following assessment by one of the certification bodies need to demonstrate that their procedures comply with the Standard. The scope of services for which registration is sought has to be clearly identified on the application for assessment and evidence demonstrating this scope has to be presented at the time of assessment.

Quantity surveyors are concerned with Part 1 of the standard, as they design their service for each individual appointment. The standard contains 20 articles and it is necessary in assessing a quality management system to identify how each of the articles is addressed. This poses certain difficulties, as the Standard was originally provided for the manufacturing industry and therefore requires interpretation.

Although some of the articles are not totally relevant to the quantity surveying function the majority are, and their application within the quality management system will range from the management responsibility of setting the policy, through the procedures for dealing with incoming and outgoing documents, to requirements with regard to auditing and training.

Once the practice is registered as a firm of assessed capability, regular audits will be carried out by the certification body to ensure ongoing compliance with all the requirements of the quality management system and continuing compliance with the Standard.

A more recent development in quality management standards, and one that will be of increasing importance to UK industry and commerce (including the construction industry), is the concept of Total Quality Management (BS 7850). This addresses the management techniques used throughout an organization to improve quality and maximize effectiveness and efficiency, whereas BS 5750 deals with the actual quality management system used.

Quality of documentation

While quality management systems set procedures to be followed to ensure quality of service provided, another aspect that is worth considering is the quality of documentation leaving the office, as this is the chief means of communication with other organizations. There will be correspondence in the form of letter or reports issued on every project, from the first letter of instruction to a final letter sending in an account for fees. The science and art of letter and report writing constitute a subject well worth studying, but only a few points can be mentioned here.

The object of writing is to convey the ideas of one person to the mind of another, who is not present to be addressed verbally, and at the same time to make a permanent record of the communications. The writer must convey by words alone both the emphasis required and the tone in which the letter or report is written. Words and phrases must, therefore, be very carefully chosen.

Without going into the subject too deeply, a few suggestions may be made:

- Be quite sure that the points made are clear.
- Be as brief and straightforward as possible and do not use two words where one will do.
- Start a new paragraph with each new point.
- If a long letter develops, consider whether it is not better to put the matter in the form of a schedule or report, with only a short covering letter.
- Be sure to write with the reader in mind. Do not use technical terms when writing to a non-technical client.
- Avoid commercial clichés, journalese, Americanisms and slang.
- Avoid spelling mistakes and bad grammar. They give a poor opinion to an educated reader.
- Avoid the impersonal.

Time and cost management

In the increasingly competitive fee and tender market the management of time and cost is of significant importance. There is a need for staff to be managed in as efficient a way as possible. To assist in this, detailed records of time expended and thereby costs incurred need to be monitored on a regular basis.

Resource allocation

Resources need to be allocated to specific tasks on one or more project depending upon their size and complexity. Teams may be assembled for document preparation or post-contract administration on large projects. The overall resource allocation needs to be regularly monitored. Certain work will be indeterminate or require immediate attention, which can cause problems in overall planning. Agreement of design and documentation production programmes with members of the design team can assist in preplanning the timing of resource requirements.

Individual time management

Each member of staff needs to manage his or her own time as efficiently as possible. This is particularly important when they are involved on a number of projects at any one time. Effective time management, although achievable in theory, becomes more of a problem in practice.

Individual tasks should be prioritized, with the most urgent being addressed first (not the most straightforward). Where deadlines are critical, a time-scale should be applied to each task to provide a target for completion.

Staff time records

It is also necessary to plan and coordinate the individual staff members' time. This information can be presented as a bar chart for each member of staff. It needs to take staff holidays into account, to be revised at regular intervals, and to be reviewed weekly.

Whether or not an organization has a full job costing system, it is necessary that detailed records be kept of the time worked by each member of staff. See the sample weekly time sheet in Fig. 5.2. This may be required as a basis on which to build up an account for fees for services that will be charged for on a time basis, or to establish the cost for each particular project. Such costs will be used primarily to establish whether a particular project is making a profit or a loss, but may also be used to estimate a fee to be quoted for similar future work.

Each member of staff needs to keep a diary in which to record not just movements, meetings, and other matters but also time spent on each project. It is essential that entries are made on a daily basis and entered up before leaving the office. In addition to identifying the project worked on, notes of specific activities, such as taking off and valuation, need to be identified in order for time sheets to be comprehensively completed.

Progress charts

The keeping of charts showing the work in the office and its progress is useful for management purposes. It can also identify future commitments in workload, although these may only be tentative. The form of a progress chart may be similar to that used for construction works, different jobs taking the place of the stages in a building contract.

Weekly Time Sheet   Name   Grade   Location   Staff No.   Week Ending

Job Description	MON		TUES		WED		THUR		FRI		SAT		SUN		TOTAL		CODE		
	BASIC	O/T	BASIC	O/T	BASIC	O/T	BASIC	O/T	BASIC	O/T	BASIC	O/T	BASIC	O/T	BASIC	O/T	JOB NO.	WORK CODE	CH/NC

Office Administration
Training/Study Leave
Public Holidays
Annual Holidays
Sick
Others (Specify)
TOTAL

FOR ANALYSIS OF WORK CODES SEE BACK OF THIS DOCUMENT

Signed (Staff)   Approved (Partner/Associate)

WORK CODES

CODE	DESCRIPTION
A	PRE CONTRACT ESTIMATES AND COST CONTROL
B1	TENDER AND CONTRACT DOCUMENTATION INCLUDING REPORTS
B2	NEGOTIATIONS
C1	POST CONTRACT – VALUATIONS AND COST REPORTS
C2	POST CONTRACT – FINAL ACCOUNTS
D	CLAIMS
E	OTHER QS SERVICES eg Liquidations, attendance at additional meetings, etc
F	PROJECT MANAGEMENT
H	FACILITIES MANAGEMENT
I	OTHER eg Insurance valuations, expert witness work, arbitration, etc

Fig. 5.2 Sample weekly time sheet.

The progress chart may also be useful in helping to decide whether to look for additional staff or whether there is an under-utilization of staff for which work in the long term will need to be forthcoming.

Education and training

The methods of entry into the profession are described in the leaflets published by the RICS.

Whether staff are selected from school-leavers, who will then attend a local college or university on a day-release basis, or directly from an undergraduate university degree course, will depend upon the practice's preference. There are advantages in both cases. Professional practice is really only something that can be learned by experience, and this argument favours the school-leaver. However, intensive study of theoretical knowledge will provide the young surveyor with a wider view of the profession. Approximately 80% of new members to the profession currently choose a sandwich or full-time course. In either case the surveyors of the future are likely to undertake some type of RICS exempting course, and professional associate membership of the Institution will be via the assessment of professional competence (APC).

The underlying objective of academic study, therefore, should be to develop in the student an understanding of the principles and concepts of the process of quantity surveying. Although this should contain some technical training, nowadays this responsibility is largely left to the professional offices and contracting organizations. Recognition as a qualified surveyor will still be in the hands of the Institution.

In 1993 the RICS published a Code of Practice and Training Agreement for the Assessment of Professional Competence whereby employers can become registered with the RICS as approved trainers. This can be achieved by adopting the model structured training scheme, or by their own in-house scheme being approved. In these circumstances the training period for candidates may be reduced from 33 to 24 months.

Upon qualifying, the surveyor should not consider this to be the end of studies. Regular continuing professional development (CPD) must be undertaken as a requirement of RICS membership. This may mean attending courses or further study for specialization in a particular aspect of the profession, such as management. It may also incorporate postgraduate courses and research.

Finance and accounts

The subject of finance and accountancy is now part of most undergraduate surveying courses. It is a practical form of economics and as such is important to the quantity surveyor in general. It is also useful to have an elementary understanding of the subject. Larger practices and companies will have finance and accounts departments with specialist accountancy staff, whereas within smaller organizations all financial matters are dealt with by senior management.

The accounts

The primary purpose of keeping accounts is to provide a record of all the financial transactions of the business, and to establish whether or not the business is making a profit. The accounts will also be used:

- in determining the distributions to be made to equity holders;
- in determining the partners' or company's tax liabilities;
- to support an application to a bank for funding;
- to determine the value of the business in the event of a sale;
- as a proof of financial stability to clients and suppliers.

All limited companies are required under the Companies Acts to produce accounts and to file them annually with the Registrar of Companies in order that they are available for inspection by any interested party.

The principal accounting statements are the profit and loss account and the balance sheet.

Profit and loss account

The profit and loss account records the results of the business's trading income and expenditure over a period of time. For a surveying practice, income will represent fees receivable for the supply of surveying services; expenditure is likely to include such items as salaries, rent and insurance. After adjustments have been made for accruals (revenue earned or expenses incurred that has not been paid or received) and prepayments (advance payments for goods or services not yet provided), an excess of income over expenditure indicates that a profit has been made. The reverse would indicate a loss.

The preparation of the profit and loss account will enable the business to:

- compare actual performance against budget;
- analyse the performance of different sections within the business;
- assist in forecasting future performance;
- compare performance against other businesses;
- calculate the amount of tax due.

An example of a simple profit and loss account is shown in Fig. 5.3.

Balance sheet

The balance sheet (Fig. 5.4) gives a statement of a business's assets and liabilities as at a particular date. The balance sheet will include all or most of the following:

- *fixed assets*: those assets held for long-term use by the business, including intangible assets;
- *current assets*: those assets held as part of the business's working capital;
- *liabilities*: amounts owed by the business to suppliers and banks;
- *owner's capital*: shareholders' funds (issued share capital plus reserves) in a limited company, or the partners' capital accounts in a partnership.

Profit and loss account for the two months to 28 February 1993

	£	£
Income		
Fees received		300 000
Expenditure		
Salaries	150 000	
Rent	50 000	
Others	30 000	
Depreciation	20 000	250 000
Profit for the period		£50 000

Fig. 5.3 Example profit and loss account.

```
Balance sheet as at 28 February 1993

                                           £                    £

Fixed assets
  Fixtures and fittings                                      200 000
  Less: Depreciation                                         20 000
                                                             180 000

Current assets
  Fees receivable                      60 000
  Cash at bank                         60 000
                                      120 000

Current liabilities                    50 000
Net current assets (working capital)                         70 000

Net assets                                                  £250 000

Capital                                                     200 000
Retained profits                                             50 000

                                                            £250 000
```

Fig. 5.4 Example balance sheet.

The various types of asset and liability accounts are considered in more detail below.

Assets
The term 'assets' covers the following:

- *intangible assets*, which include goodwill, trademarks and licensing agreements, usually at original cost less any subsequent write-offs;
- *fixed assets*, which include land and buildings, fixtures and fittings, equipment and motor vehicles, shown at cost or valuation less accumulated depreciation;
- *current assets*, which are held at the lower cost or net realizable value. Current assets include cash, stock, work in progress, debtors and accruals in respect of payments made in advance.

Depreciation, referred to above, records the loss of value in an asset resulting from usage or age. Depreciation is charged as an expense to the profit and loss account, but is disallowed and therefore added back, for tax purposes. Depreciation is recorded as a credit in the balance sheet, reducing the carrying value of the firm or company's fixed assets.

Liabilities
Liabilities include amounts owing for goods and services supplied to the business and amounts due in respect of loans received. Strictly it also includes

amounts owed to the business's owners: the business's partners or shareholders. Note that contingent liabilities do not form part of the total liabilities, but will appear in the form of a note on the balance sheet as supplementary information.

Capital

Sources of capital may include proprietors' or partners' capital or, for a limited company, proceeds from shares issued. Capital is required to fund the start-up and subsequent operation of the business for the period prior to that period in which sufficient funds are received as payment for work undertaken by the business.

The example in Fig. 5.4 identifies an initial capital investment of £200 000; however, this investment is soon represented not by the cash invested but by various other assets and liabilities, as shown.

To illustrate the movement of cash in terms of receipts and payments a simple example of a summarized bank account statement is included in Fig. 5.5. This statement is produced periodically, usually monthly, and is the source of the cash postings to the other books of account.

Finance

There are several ways in which a business can supplement its finances. The most common way is by borrowing from a bank on an overdraft facility. The lender will be interested in securing both the repayment of the capital lent and the interest accruing on the loan. The lender will therefore require copies of the business's profit and loss account, balance sheet and details of its projected cash flow.

Cash forecasting and budgeting

It is necessary for a business to predict how well it is likely to perform in financial terms in the future. Budgets are therefore prepared, usually on a annual basis, based on projected income and expenditure. Once a business is established, future projections can be based to a certain extent on the previous year's results.

As mentioned above, in the event that it is the intention to borrow money from

Receipts	£	Payments	£
Capital introduced	200 000	Salaries	150 000
Fees received	240 000	Rent	50 000
		Other expenses	30 000
		Fixtures and fittings	150 000
		Balance carried forward	60 000
	£440 000		£440 000

Fig. 5.5 Bank account summary statement.

a bank then the bank is likely to request a cash-flow forecast for the next six or twelve months. The preparation of a cash-flow forecast is a relatively easy process and in practice a computerized spreadsheet package or accounting software will be used to project the likely phasing of receipts and payments.

The example cash-flow forecast in Fig. 5.6 illustrates the starting-up of a professional business. It identifies the initial introduction of capital, the borrowing facility requested and the projected effect of expenditure and receipts over the period. It can be seen from the forecast that an additional £10 000 of funding will be required in April and a further (£25 000–£10 000) = £15 000 in May.

This cash flow is typical of a business start-up, when substantial sums are spent in advance of income being received. Provided the business is run profitably the outflow should be reversed before too long.

Books of account

The underlying books of account are likely to comprise the general ledger (which will include all general items such as salaries and rents, and totals from the subsidiary ledgers such as the sales ledger (fees or other income receivable), the

	January	February	March	April	May	June
Capital introduced	200 000	—	—	—	—	—
Fees received	60 000	180 000	50 000	110 000	100 000	200 000
Asset sales	—	—	—	—	—	—
Receipts	260 000	180 000	50 000	110 000	100 000	200 000
Salaries	75 000	75 000	75 000	75 000	75 000	75 000
Rent	25 000	25 000	25 000	25 000	25 000	25 000
Equipment	100 000	50 000	20 000	30 000	—	—
Others	15 000	15 000	15 000	15 000	15 000	15 000
Payments	215 000	165 000	135 000	145 000	115 000	115 000
Movement in cash	45 000	15 000	(85 000)	(35 000)	(15 000)	85 000
Balance brought forward	—	45 000	60 000	(25 000)	(60 000)	(75 000)
Balance carried forward	45 000	60 000	(25 000)	(60 000)	(75 000)	10 000
Borrowing facility	50 000	50 000	50 000	50 000	50 000	50 000
Additional requirement	—	—	—	10 000	25 000	—

Fig. 5.6   Example cash-flow forecast.

bought ledger (accounts payable), the cash book (a record of the bank transactions) and the petty cash account. Other books, such as fee and expenses books, may also be kept.

In addition to the books of account, businesses must retain vouchers such as receipts, invoices, fee accounts and bank statements to support the accounting records. These are required by businesses' auditors and for VAT purposes.

Computerization

Accounting functions have become less time-consuming through the use of computers for the regular and routine entries and calculations that are necessary. Entries can be allocated to different accounts, and up-to-date information can be retrieved quickly and efficiently in a variety of formats to meet particular needs. This greatly assists in the financial management of a business.

Annual accounts/auditing

At the end of a business's financial year a set of 'end of year' accounts are prepared bringing together all the previous year's financial information in the form of a profit and loss account and balance sheet as described earlier.

In most circumstances accounts will be audited by an independent accountant; indeed this is a requirement for all larger limited companies under the Companies Acts. Audited accounts will carry more authority with the Inspector of Taxes and are also useful to prove to third parties, including prospective clients, that the financial status of the business has been independently scrutinized.

Bibliography

Ashworth, A. *The Education and Training of Surveyors.* Construction Information File, Chartered Institute of Building, Summer 1994.

Barrett, P. *Profitable Practice Management for the Construction Professional.* Spon, 1993.

Bevan, O. *Marketing and Property People.* Macmillan, 1991.

Brandon, P.S. (ed.) *Quantity Surveying Techniques: New Directions.* Blackwell Scientific Publications, 1992.

Brody, E.W. *Professional Practice Development: Meeting the Competitive Challenge.* Praeger, 1990.

Chalkley, R. *Professional Conduct: A Handbook for Chartered Surveyors.* RICS Books, 1990.

College of Estate Management *Marketing and Chartered Quantity Surveyors.* College of Estate management, 1992.

Dann, C. 'The responsibility of a profession'. *Chartered Quantity Surveyor*, December 1983.

Davis, Langdon and Everest *QS 2000: The Future of the Chartered Quantity Surveyor.* The Royal Institution of Chartered Surveyors, 1991.

Flanagan, R. *Risk Management and Construction.* Blackwell Scientific Publications, 1992.

Hill, B. 'Promotion for the QS'. *Chartered Quantity Surveyor*, March 1985.

ICAEW *Auditor's Report on Financial Statements.* Chartac Books, 1989.

Land, H. and Collett, P. *The Chartered Surveyors Assessment of Professional Competency.* Owlion Audio Programme, 1992.

Langford, D. and Male, S. *Strategic Management in Construction.* Gower, 1991.

Male, D. 'The appointment of quantity surveyors in a changing world'. *Chartered Quantity Surveyor*, November 1982.

Male, S. 'Professional authority, power and emerging forms of "profession" in quantity surveying'. *Construction Management and Economics*, Spring 1990.

Parris, J. *Companies for Construction Professionals*. Blackwell Scientific Publications, 1987.

RICS *A Study of Quantity Surveying and Client Demand*. The Royal Institution of Chartered Surveyors, 1984.

RICS *Royal Charter and Bye-Laws*. RICS Books, 1993.

RICS *Rules and Guide to the Assessment of Professional Competence*. Quantity Surveyors Division, the Royal Institution of Chartered Surveyors, 1992.

RICS *The Chartered Surveyors' Rule Book*. The Royal Institution of Chartered Surveyors, 1993.

Thompson, F.M.L. *Chartered Surveyors: The Growth of a Profession*. Routledge and Kegan Paul, 1968.

Tracey, J.A. *How to Read a Financial Report*. Wiley, 1989.

Walker, A. and Flanagan, R. *Property and Construction in Asia Pacific*. Blackwell Scientific Publications, 1991.

Wilson, A. *Practice Development for Professional Firms*. McGraw-Hill, 1984.

Wood, F. *Business Accounting*. Longman, 1989.

Chapter 6

The quantity surveyor and the law

Introduction

The purpose of this chapter is to describe in general terms how the law affects the quantity surveyor in practice, concentrating on the form of agreement between the quantity surveyor and the client, the impact of the demand for collateral warranties and performance bonds, the requirement to maintain professional indemnity insurance cover and the Employment Acts as they relate to the employment of staff.

The wider issue of construction law is a complex and ever-changing area, the detail of which falls outside the scope of this book, it being adequately covered in the specialist texts on the subject. Reference is however made in general terms within Chapters 8–11 to the areas of legal significance that are worthy of note by quantity surveyors whether in private practice or in contracting organizations.

The quantity surveyor and the client

The quantity surveyor in private practice provides professional services for a client. The legal relationship existing between them is therefore, or ought to be, a contract for services. The nature of this contract controls the respective rights and obligations of the parties and, so far as the quantity surveyor is concerned, determines the duties to be performed, powers and remuneration in respect of the particular work undertaken.

Agreement for appointment

There is no legal requirement that the agreement or contract between the surveyor and the client should consist of a formal document nor even, indeed, that it be in writing. Nevertheless, given that it may be crucial to establish the precise nature of the relationship in the event of a dispute, it is desirable that the understanding reached be confirmed in writing. If differences subsequently arise and proceed to litigation, the court will be faced with the problem of ascertaining the true intentions of the parties from the available evidence. Clearly a written record will, in these circumstances, be a great deal more persuasive than the possibly disputed recall of the contending parties.

As mentioned, no particular formality is required and both the existence and nature of the agreement can be established by an exchange of correspondence, by the use of a standard form of appointment such as the Form of Agreement,

Terms and Conditions for the Appointment of a Quantity Surveyor published by the RICS, or other non-standard form. Whatever form is adopted, the need to ensure that a valid, clear-cut, comprehensive and adequately evidenced contract exists, cannot be overstressed.

Professionals can generally find themselves in great difficulty if they undertake work relying upon an incomplete agreement, in which important items remain to be settled. Given the real pressures to acquire business, it is easy to succumb to this temptation without adequately weighing the risks involved. The law, however, does not recognize the validity of a contract to make a contract and, where any essential element is left for later negotiation, the existing arrangements are unlikely to be recognized as a binding agreement.

In *Courtney & Fairbairn Ltd* v. *Tolaini Bros (Hotels) Ltd* (1975) 1 AER 716; 2 BLR 100, the defendants, wishing to develop a site, agreed with the plaintiffs, a firm of contractors that if a satisfactory source of finance could be found, they would award the contract for the work to the plaintiffs. No price was fixed for the work but the defendants agreed that they would instruct their quantity surveyor to negotiate fair and reasonable contract sums for the work. Suitable finance was introduced, and the quantity surveyor instructed as agreed. The quantity surveyor was unable, in the event, to negotiate acceptable prices, resulting in the contract being let elsewhere. The plaintiffs sued for damages, claiming that an enforceable contract had come into being. It was held that the price was a fundamental element in a construction contract and the absence of agreement in that regard rendered the agreement too uncertain to enforce.

The foregoing case was, of course, not directly concerned with the provision of professional services but the legal principle illustrated is of general application.

A persuasive incentive to take care arises from the fact that the ability to recover payment for work done will usually depend on the existence of an appropriate contract. Performance alone does not automatically confer a right to remuneration; although where a benefit is conferred, the court will normally require the beneficiary to make some recompense, possibly by way of a *quantum meruit* payment, meaning 'as much as is deserved'.

In *William Lacey Ltd* v. *Davies* (1957) 2 AER 712, the plaintiff performed certain preliminary work for the defendant, connected with the proposed rebuilding of war-damaged premises, in the expectation of being awarded the contract for the work. The defendant subsequently decided to place the contract elsewhere, and eventually sold the site without rebuilding. The plaintiff sued for payment for work already done. In this case it was held that in respect of the work done no contract had ever come into existence but, nevertheless, as payment for the work had always been in the contemplation of the parties, an entitlement to some payment on a *quantum meruit* arose.

In this context it is reassuring, from the quantity surveyor's point of view, to note that, where professional services are provided, there is a general presumption that payment was intended. In *H.M. Key & Partners* v. *M.S. Gourgey & Others* (1984) I CI.D–02–26, it was said: 'The ordinary presumption is that a professional man does not expect to go unpaid for his services. Before it can be held that he is not to be remunerated there must be an unequivocal and legally enforceable agreement that he will not make a charge'. However, while some recovery of fees may be possible without the formation of a binding contract, the lack of such an agreement enhances the possibility of disputes and litigation. Moreover, if the basis of enforced payment is to be *quantum meruit*, there is no guarantee that the court's evaluation of the services provided will correspond with the practitioner's expectations.

Given that all relevant terms are settled and agreed, and incorporated in a formal contract, the intentions of the parties may still be frustrated by a failure to express the terms clearly.

In *Bushwall Properties Ltd* v. *Vortex Properties Ltd* (1976) 2 AER 283, a contract for the transfer of a substantial site provided for staged payments and corresponding partial legal completions. At each such completion 'a proportionate portion of the land' was to be transferred to the buyer. A dispute arose as to the meaning of this phrase. It was held that in the circumstances no certain meaning could be attributed to the phrase; that this represented a substantial element in the contract; hence the entire agreement was too vague to enforce.

However, if a valid contract exists in unambiguous terms, the court will enforce it. It is therefore vital to ensure that the terms are not merely clear but do in fact represent the true understanding of the parties, both at the outset and as the work progresses. When the actual work is in hand with all attendant pressures it is all too easy to overlook the fact that the obligations undertaken and remuneration involved are controlled by the contract terms. Departure from or misunderstanding of the original intentions, unless covered by suitable amendments of those terms, may have very undesirable consequences, as the following case illustrates.

In *Gilbert & Partners* v. *R. Knight* (1968) 2 AER 248; 4 BLR 9, the plaintiffs, a firm of quantity surveyors, agreed for a fee of £30 to arrange tenders, obtain consents for, settle accounts and supervise certain alterations to a dwelling house on behalf of the defendant. Initially work to the value of some £600 was envisaged but in the course of the alterations the defendant changed her mind and ordered additional work. In the end, work valued at almost four times the amount originally intended was carried out; the plaintiffs continued to supervise throughout and then submitted a bill for £135. This was met with a claim that a fee of £30 only had been agreed. It was held that the original agreement was for an all-in fee covering all work to be done; the plaintiff was entitled to only £30.

The moral is clear: avoidance of difficulty and financial loss is best ensured by accepting engagement only on precise, mutually agreed and recorded terms, setting out unequivocally what the quantity surveyor is expected to do and what the payment is to be for so doing. If, as often happens, circumstances dictate development and expansion of the initial obligations, the changes must be covered by fresh, legally enforceable agreements. Oral transactions, relating to either the original agreement or later amendment of it, should always be recorded and confirmed in writing. This is more than an elementary precaution, for it should be borne in mind that what is known as the parol evidence rule will normally preclude any variation of an apparently complete and enforceable existing written contract, by evidence of contrary or additional oral agreement.

The agreement for appointment of a quantity surveyor, whether a standard or non-standard document, will encompass certain general provisions including the following:

- form of agreement/particulars of appointment
- scope of services to be provided
- fee details
- payment procedures
- professional indemnity insurance requirements
- assignment
- suspension
- copyright
- duty of care
- dispute procedures.

Other provisions, such as limitation of the quantity surveyor's authority, communications and duration of appointment, might be included.

The agreement can be executed either as a simple contract or as a deed. There are important differences, two of which are the most significant as far as the quantity surveyor is concerned: the need for consideration and the limitation period.

In a simple contract there must be consideration (benefit accruing to one party or detriment to the other, most commonly payment of money, provision of goods or performance of work), and the period in which an action for breach of contract can be brought by one party against the other is limited to six years. In a speciality contract (a contract executed as a deed), however, there is no need for consideration and the limitation period is 12 years. The significance of the latter is that a quantity surveyor who enters into an agreement as a deed doubles the period of exposure to actions for breach of contract.

Responsibility for appointment

What has been said so far assumes that the quantity surveyor's appointment arose from direct contact with the client. Additional problems may occur where, as may be the case, the appointment arises indirectly from approaches by the retained architect or project manager. In such cases the power to appoint on behalf of the client may subsequently be called into question. There is no general answer to this problem; the actual position will depend on the express and implied terms of the other consultant's contract with the client. If they have express power to appoint, then, of course, no problem arises and the appointment is as valid as if made by the client in person. Reliance on their possessing an implied power to appoint would, however, be very unwise. It seems clear that the courts do not recognize any general power of appointment or delegation as inherent in an architect's or other consultant's contract with a client.

In *Moresk Cleaners Ltd.* v. *T.H. Hicks* (1966) 4 BLR 50, the Official Referee stated bluntly that 'The architect has no power whatever to delegate his duty to anyone else'. That case concerned the delegation of design work but it would seem equally applicable to other unauthorized appointments.

This absence of a general implied power to appoint does not preclude the possibility that in particular circumstances it may be held to exist. There is, for example, some rather dated authority for the proposition that, where tenders are to be invited on a bill-of-quantities basis, such implied authority may be present.

Potential difficulties in the matter can be easily avoided by the simple expedient of ensuring that, where the employment of the quantity surveyor is negotiated by another consultant, the terms of the appointment are conveyed in writing to the client and the client's acceptance thereof similarly secured. Ratification by the client will then have overcome any deficiencies in the consultant's authority.

Responsibility for payment of fees

Where an effective contract exists between the quantity surveyor and the client, provision will undoubtedly be contained relating to the payment of the professional fees involved. No real difficulty should be experienced where the appointment has been made by a duly authorized agent, or the client has ratified an appointment purported to have been made on their behalf. The position where no valid agreement exists has already been mentioned, and it was suggested that even in such unfortunate circumstances some remuneration, probably by way of a *quantum meruit*, will usually be forthcoming.

If the authorized agent, however, has made the appointment in excess of their powers, the quantity surveyor will have to look elsewhere than to the client for payment. In such circumstances it will usually be possible to bring an action against the agent personally. Such a claim would normally lie either for warranty of authority, where the agent has misunderstood or exceeded the authority granted by the client, or had not actually purported to act on the client's behalf. The existence of a legal remedy is nevertheless of doubtful consolation if the debtor is unable to pay.

If any reservations are entertained as to the financial standing of a potential client, the surveyor must make enquiry as practicable, perhaps by taking up bank references, and then trust to commercial judgement. In this connection it is vital to keep clearly in mind, and ensure that the documentation accurately reflects, the true identity of the client. This may seem too obvious to mention but misunderstandings can and do occur, particularly in dealings with smaller companies controlled and dominated by an individual. It is easy to confuse the individual acting on behalf of a company and acting in a personal capacity. The unhappy result may be dependence for payment on a company of doubtful solvency, having imagined that one was acting for an individual of undoubted substance. Finally, it is chastening to reflect that monies owed in respect of professional fees are in no way preferred in the event of insolvency.

Amount and method of payment

Actual fees and fee rates are now a matter for negotiation, although recommended scales exist. Entitlement depends on the terms of the agreement under which the services are provided and any negotiations are constrained by practical rather than legal considerations. Where, however, the work involves advising on matters connected with litigation or arbitration, say in respect of claims, it is not permissible to link the fee to the amount recovered. In *J. Pickering* v. *Sogex Services Ltd* (1982) 20 BLR 66, arrangements of that nature were said to savour of champerty – that is, trafficking in litigation – and as such to be unenforceable as contrary to public policy.

It is prudent, where possible, to make provision for the payment of fees by instalments at appropriate intervals. The payment of a series of smaller amounts while services are being provided tends to be more readily accepted than the settlement of a substantial bill when the work has been completed. Moreover, failure to pay on time may be a useful guide to the state of the client's finances. If the worst happens and the client is rendered insolvent, there is some comfort in being an ordinary creditor for only the balance and not the entirety of the fee is involved. Failure to pay on time should be treated seriously and, if necessary, legal steps for recovery should be taken promptly. In these matters patience and understanding are more likely to lead to disappointment than to reward.

Negligence

Broadly speaking, where the law is concerned, negligence usually consists either of a careless course of conduct or such conduct, coupled with further circumstances, sufficient to transform it into the tort of negligence itself. As stated earlier the extent and nature of the duties owed to the client by the quantity surveyor, as well as the powers and authority granted to the client, will be determined by the contract for services between them.

It has always been implied into a professional engagement that the professional person will perform duties thereunder with due skill and care, and this requirement is reiterated by provisions in the Supply of Goods and Services Act 1982. Lack of care in discharging contractual duties is, and always has been, an actionable breach of contract.

As late as the mid-1960s, when it was so held in *Bagot* v. *Stevens Scanlan & Co.* (1964) 3 AER 577, the existence of a contractual link between the parties was believed to confine liability to that existing in contract and to exclude any additional liability in tort. Since then the position has gradually changed, and the courts appeared to recognize virtually concurrent liability in both contract and tort. Thus an aggrieved contracting party was able to sue the other contracting party or parties both in contract and tort. This was illustrated by the decisions in *Midland Bank Trust Co. Ltd* v. *Hett, Stubbs & Kemp* (1978) 2 AER 571 and, more immediately relevant to the construction industry, in *Batty* v. *Metropolitan Property Realisations Ltd* (1978) 7 BLR 1.

Liability was also considered to exist independently, where there is no contractual link between the parties, enabling a third party to sue in the tort of negligence. A plaintiff suing in negligence must show that:

- the defendant had a duty of care to the plaintiff, and
- the defendant was in breach of that duty, and
- as a result of the breach the plaintiff suffered damage of the kind that is recoverable.

In the first place, the plaintiff would try to show that the defendant owed a duty of care. From the principles established in *Donoghue* v. *Stevenson* (1932) AER 1 (the celebrated 'snail in the bottle' case), the courts tended to find the presence of a duty of care in an ever-increasing number of circumstances.

The tentacles of the tort of negligence even extended well beyond normal commercial relationships, at least so far as the professional person was concerned. Since the well-known case of *Hedley Byrne & Co Ltd* v. *Heller & Partners Ltd* (1963) 2 AER 575, any negligent statement or advice, even if given gratuitously, seemed in certain circumstances to afford grounds for action.

In more recent times in the case of *Junior Books Ltd* v. *The Veitchi Co. Ltd* (1982) 21 BLR 66 the House of Lords held that a specialist flooring subcontractor was liable in negligence for defective flooring to the employer with whom the sub-contractor had no contractual relationship. Almost immediately, however, the courts began to retreat from the position by means of a long string of cases which culminated in *Murphy* v. *Brentwood District Council* (1990) 50 BLR 1 which, among other things, overturned the 12-year-old decision in *Anns* v. *London Borough of Merton* (1978) 5 BLR 1.

The tortious liability for negligence is therefore reduced, which in itself has led to the growth in the use of collateral warranties.

There is little case law concerning the negligence of a quantity surveyor; however, it is always advisable to limit the risk, and the best safeguard is discretion and a reluctance to express opinions or proffer advice on professional matters, unless one is reasonably acquainted with the relevant facts and has had the opportunity to give them proper consideration.

Death of the surveyor

Whether the liability to carry out a contract passes to the representatives of a deceased person depends on whether the contract is a personal one: that is, one in which the other party relied on the 'individual skill, competency or other personal qualification' of the deceased. This is a matter to be decided in each particular case.

In the case of a surveyor with no partner the appointment must be regarded as personal, and the executors could not nominate an assistant to carry on with business, except in so far as the respective clients agreed. With a firm of two or more partners the appointment may be that of the firm, in which case the death of one partner would not affect existing contracts. But the appointment may be of one individual partner, as an arbitrator for example, where another partner in the firm could not take over, even though he or she may be entitled to a share of the profits earned by the partner in the arbitration.

The fact that a contract between a quantity surveyor and the client is a personal contract, if such be the case, does not mean that the quantity surveyor must personally carry out all the work under the contract, unless it is obvious from the nature of the contract, for example a contract to act as arbitrator, that the quantity surveyor must act personally in all matters. In other cases, such as the preparation of a bill of quantities and general duties, the quantity surveyor may make use of the skill and labour of others, but takes ultimate responsibility for the accuracy of the work.

Death of the client

The rule referred to in the previous paragraph as to a contract being personal applies equally in the case of the death of the client. Here the contract is unlikely to be a personal one, and the executors of the client must discharge the client's liabilities under the building contract and for the fees of the professional people employed. The fact that the appointment of the surveyor was a personal one will not be material in the case of the death of the client.

Collateral warranties

A collateral warranty (or duty of care agreement) is a contract that operates alongside another contract and is subsidiary to it. In its simplest form it provides a contractual undertaking to exercise due skill and care in the performance of certain duties, which are the subject of a separate contract (an example of which would be a warranty given by a quantity surveyor to a funding institution to exercise due skill and care when performing professional services under a separate agreement between the quantity surveyor and the client).

The purpose of a collateral warranty is to enable the beneficiary to sue the party giving the warranty for breach of contract if the warrantor fails to exercise the requisite level of skill and care in the performance of the duties.

It used to be the view that such an agreement was not very important because it merely stated in contractual terms the duties that were owed by the quantity surveyor to a third party in tort. That view is no longer tenable.

There is a fundamental contract principle that only the parties to a contract have any rights or duties under that contract. The principle is called 'privity of contract'. For example, in a contract between a client and quantity surveyor, each has rights and duties to the other. The quantity surveyor has a duty to carry out specific duties but has no duty to any third party. That would be the case even if the contract stated that there was such a duty.

At one time the third party would have been able to overcome this kind of problem by suing in the tort of negligence if there was no contractual relationship. This is no longer the case, however. As stated earlier in connection with negligence the law in this regard has changed over recent years, culminating in the case of *Murphy* v. *Brentwood District Council* (1990) 50 BLR 1.

As a result of the fundamental changes in the law, collateral warranties are now of significant importance to clients and have proliferated in recent years. It is now common for contractors, nominated and domestic subcontractors and suppliers and all the consultants to execute a collateral warranty in favour of the client, the company providing the finance for the project and/or prospective purchasers/tenants.

There are standard forms of warranty to be given to funders and purchasers or tenants published by the BPF which have been approved by the RIBA, RICS, ACE (Association of Consulting Engineers) and the BPF. There are also a great many other forms of warranty in circulation, some of which have been especially drafted by solicitors with a greater or lesser experience of the construction profession and the construction industry generally.

The party requiring collateral warranties in connection with a project would probably gain maximum benefit from those provided by the design consultants. It is more likely that a claim under a warranty will be related to design matters. It is common however for warranties to be requested from all consultants on a project including the quantity surveyor. It is important to ensure that all consultants enter into a warranty on the same terms (with the exception of clauses relating to the selection of materials: see below).

The following are specific issues that should be considered by quantity surveyors when called upon to sign a warranty.

Relationship to terms of appointment

Prior to agreeing any warranty terms it is essential to have full written terms of appointment as described earlier. Unless the duties to be undertaken by the quantity surveyor are fully detailed, the standard warranty term that 'all reasonable skill and care be taken in the performance of those duties' leaves it open to argument as to the definition of such duties if a claim is brought under the warranty in the future.

The warranty should refer specifically to the terms of appointment, and it is important that the terms and conditions of the warranty are no more onerous than those contained in the appointment.

Materials

The standard forms of warranty include provisions regarding taking reasonable skill and care to ensure that certain materials are not specified. This clause should be deleted from quantity surveyors' warranties as it relates primarily to design consultants.

Assignment

The warranty is likely to make provision for the warrantee to be able to assign the benefits to other parties. The more restrictive the assignment clause the better as far as the quantity surveyor's potential liability is concerned. Much of the value of the warranty is the ability to assign, and therefore if a quantity surveyor agrees to an assignment clause it should be limited in terms of number of assignments and time-scale: for example, assignment only once within a limit of three years subject to the quantity surveyor's consent.

Professional indemnity

The warrantee will be concerned with the level of professional indemnity insurance cover carried by the quantity surveyor, and the required level of cover will be stated in the warranty. It is important that all warranties are passed to the insurers before being signed or the insured could be at risk.

A requirement to maintain professional indemnity insurance cover at a level for a specific number of years is impractical; a requirement to maintain cover should be limited to using best endeavours to maintain cover as long as it is available at commercially reasonable rates.

Complete records should be kept of all warranties given, as it is necessary to disclose these annually to the insurers at the time of renewal of the policy.

Execution

The essential differences between a simple contract and a deed have been highlighted earlier. If the quantity surveyor is requested to enter into a warranty agreement as a deed whereas the terms of appointment are executed as a simple contract the quantity surveyor's period of liability to the third party will be twice as long.

Performance bonds

There is an increasing demand for consultants to provide performance bonds, particularly on major projects. Although it is a concern that clients consider it necessary to require such bonds, in the pursuit of work the quantity surveyor may not be in a position to object.

The conditions of the bond are likely to be similar to those required from a contractor. The value is calculated as a percentage of the total fee (10%, for example) and the conditions under which it can be called upon are stated. In certain instances 'on demand' bonds are being requested, whereby payment by the surety can be demanded without the need to prove breach of contract or damages incurred as a consequence.

It is therefore important to check the conditions in detail and to ascertain the cost of providing the bond prior to agreeing fee levels and terms of appointment.

Professional indemnity insurance

It has always been prudent for quantity surveyors to protect themselves against possible claims from their clients for negligence for which they may be sued. Such mistakes may not necessarily be those of a principal's own making but those of an employee. The RICS by-laws and regulations make it compulsory for practices, firms or companies to be properly insured against claims for professional negligence for specified minimum levels of indemnity. Premiums are calculated according to the limit of indemnity selected, the number of partners or directors, the type of work that is undertaken and the fee income. The policy must be no less comprehensive than the form of the RICS Professional Indemnity Collective Policy as issued by RICS Insurance Services Ltd.

The policy covers claims that are made during the period when the policy is effective, regardless of when the alleged negligence took place. Claims that occur once the policy has expired, even though the alleged event took place some time previously, will not be covered. A sole practitioner is therefore well advised to maintain such a policy for some time after retirement. Recent court cases suggest that a professional person may be held legally liable for actions for a much longer period than the normal statutory limitation period would otherwise suggest.

Contracts of employment

There are certain legal requirements relating to the employment of staff. The basic relationship between the employer and the individual employee is defined by the contract of employment. This is a starting point for determining the rights and liabilities of the parties. Although these rights originated from different statutes, they are now consolidated in the Employment Protection (Consolidation) Act 1978, though further amendments have been introduced by the Employment Acts of 1980 and 1982. An important feature of these rights is that they are not normally enforced in the courts but in industrial tribunals.

A contract of employment must be given to each employee within 14 days of commencing employment. It should cover matters regarding the conditions of employment, including hours of work, salary, holiday entitlement, sick leave, termination, and the procedures to be followed in the event of any grievance arising.

Other Acts that are worthy of note are the Sex Discrimination Act 1975, whereby a person cannot be discriminated against because of his or her sex or marital status, and the Race Relations Act 1976, which considers discrimination on the grounds of colour, ethnic or national origins or nationality. These Acts cover not only recruitment but also promotion and other non-contractual aspects of employment. The Equal Pay Act 1970 covers contractual terms and conditions of employment in addition to pay, making it unlawful for an employer to treat a woman differently because of her sex.

Regarding termination of employment, the employer must give the employee the amount of notice to which he or she is entitled under the contract of employment. This will relate to the employee's length of service up to a maximum of 12 weeks, although the contract may specify a longer period. Employees may be dismissed for acts of misconduct such as theft or physical violence, but the tribunal must be satisfied that the employer acted reasonably should a complaint be brought to them.

Sometimes a job comes to an end because the firm has no more work or because the kind of work undertaken by the employee has ceased or diminished. In these circumstances the employee will normally be entitled to redundancy payments. In order to establish a claim the employee must have at least two years' service, be between the ages of 20 and 65, and have been working for a minimum number of hours per week, depending upon the length of service.

Bibliography

The quantity surveyor and practice

Cecil, R. *Professional Liability*. Legal Studies and Services (Publishing), 1991.

Cornes, D.L. *Design Liability in the Construction Industry*. Blackwell Scientific Publications, 1994.

Dugdale, A.M. and Stanton, K.M. *Professional Negligence*. Butterworth, 1989.

Holyoak, J. *Negligence in Building Law: Cases and Commentary*. Blackwell Scientific Publications, 1992.

James, P.S. *Introduction to English Law*. Butterworth, 1989.

Jess, D.C. *Insurance of Professional Negligence Risks*. Butterworth, 1989.

RICS *Caveat Surveyor II*. RICS Books, 1990.

RICS *Caveat Surveyor: Negligence Claims Handled by RICS Insurance Services*. RICS Books, 1986.

RICS *Compulsory Professional Indemnity Regulations*. The Royal Institution of Chartered Surveyors, 1993.

RICS *Direct Professional Access to Barristers*. RICS Books, 1989.
Speaight, G. and Stone, G. *The AJ Legal Handbook*. Architectural Press, 1988.

The quantity surveyor and contracts

Bickford-Smith, S. *et al. Emden's Building Contracts and Practice*. Butterworth, 1990.
Chappell, D. *Understanding JCT Standard Building Contracts*. Spon, 1993.
Egglestone, F.N. and Walmesley, R.M. *Introduction to Contractors' All Risks Insurance*. Paperback Books, 1990.
Furmston, M.P. *Cheshire, Fifoot and Furmston's Law of Contract*. Butterworth, 1991.
Keating, D. *Law and Practice of Building Contracts*. Sweet and Maxwell, 1991.
Lloyd, H. and Rease, C. *Building Law Reports*. Longman, 1987.
NJCC *Collateral Warranties*. Guidance Note 6, National Joint Consultative Committee, 1992.
Parris, J. *Arbitration: Principles and Practice*. Blackwell Scientific Publications, 1985.
Parris, J. *The Standard Form of Building Contract: JCT 80*. Blackwell Scientific Publications, 1994.
Powell-Smith, V. and Sims, J. *Building Contract Claims*. Blackwell Scientific Publications, 1988.
Powell-Smith, V. and Furmston, M.P. *A Building Contract Casebook*. Blackwell Scientific Publications, 1990.
Powell-Smith, V. and Chappell, D. *A Building Contract Dictionary*. Legal Studies and Services (Publishing), 1991.
RICS *Arbitration*. QS Practice Pamphlet No. 4, RICS Books, 1983.
RICS *Insolvencies*. QS Practice Pamphlet No. 5, RICS Books, 1985.
Trickey, G. *Presentation and Settlement of Contractors' Claims*. Spon, 1988.
Uff, J. *Construction Law*. Sweet & Maxwell, 1991.
Wallace, I.N.D. *Construction Contracts*. Sweet & Maxwell, 1986.
Wood, R.D. *Building and Civil Engineering Claims*. Estates Gazette, 1985.

Chapter 7

Information technology

Introduction

'Information technology' is a relatively recent addition to the English language. It has its counterparts in the French *informatique* and the Russian *informatika*. For many people information technology is synonymous with new technology such as microcomputers, telecommunications, computer-controlled machines and associated equipment.

The use of computers has had a dramatic influence upon human behaviour and development during the past ten years. Computers have also had a major impact upon the profession of quantity surveying, in respect of the role and function of the quantity surveyor's professional activities. While the capability of computers and their associated software continues to increase, their relative (and real!) price decreases. Their reliability is now of a high order and their use has become easier as simplified and user-friendly procedures have become commonplace. The use of information technology (IT) has wide-ranging implications. From a social point of view, IT has changed the way in which we communicate and reach decisions, manage our work and store information.

Computer literacy

Computer literacy requires an understanding of the following two related areas of computer knowledge:

- *knowing computer capabilities and limitations*: general understanding of the organization, capabilities and limitations of the various machines (the hardware);
- *knowing how to use computers*: familiarity with the common uses or applications of computers; ease in working with prewritten software.

Additional competence is gained by mastering of the following two additional areas:

- *knowing how computer software is acquired*:general idea of how individuals and organizations develop custom-made programs and information systems;
- *understanding the computer's impact*: awareness of the impact that computers and information systems are having on people and organizations.

Hardware selection

A brief introduction to the available hardware will suffice. Readers who want to explore the matter in more detail can refer to more specialized texts or the plethora of computer journals and magazines that are available.

There are essentially three types of computer hardware. Mainframes were the early generation of computers and are still required in large organizations with extensive computer demands. Minicomputers are physically smaller than mainframes but have computing power that is rapidly approaching that of the mainframe. Microcomputers are the machines that are most likely to suit the needs of the practising quantity surveyor and few firms today will be able to exist without one. These may be in the form of a desktop machine or, where portability is required, a laptop. Their price typically starts at around £1000 for a single system. Some of the main considerations to be taken into account are as follows:

- physical dimensions
- visual display type and keyboard ergonomics
- portability
- memory capacity
- capacity for mass storage
- types of storage device
- operating system
- speed of access
- compatibility with other computers and equipment
- enhancement capabilities.

The 1980s saw the IBM (International Business Machines) PC rise from nowhere to domination of the world market in personal and business micro-computers. The Intel 8088 microprocessor was at the heart of the original IBM PC. The dominant operating system at the time was CP/M. The later machines used MS-DOS, Microsoft's operating system. Neither the Intel chip nor MS-DOS are proprietary to IBM. Many of the compatible machines or clones also adopted these two features. The first company to produce a PC-compatible was Compaq, who announced a portable PC only a year after IBM had introduced their machine. Compaq, in common with other makers of PC-compatibles, sold their machines on the strength of supplying more features as standard than IBM, while matching them in quality and undercutting them on price.

Microprocessor

The microprocessor 'chip' is the heart of the machine. The more powerful it is the more things the computer will be able to do and at a faster speed. The chips found in IBM compatibles are referred to by their numbers. 8088 and 8086 are the processors in the early PCs. The next generation used a more powerful 80286 (or 286 for short) chip and the later versions use a 80486 (486) chip and now a Pentium chip (586). Modern chips are available in a number of different speed ratings (for example, 16, 20, 25 and 33 MHz). Apple Macintosh computers use different chips and are not IBM-compatible without additional software being added.

Displays: video standards

What you see on the screen can vary depending on the graphics or video card that has been installed in the computer. It is not always WYSIWYG (what you see is what you get). The video card is a smaller subcomputer that controls everything that is sent by the computer to the screen. The resolution of detail is important, and this is dependent upon the number of dots or pixels that can be shown on a screen in a standard area: the larger the number the finer the detail that you can show on the screen. The older IBM machines could only display text (letters and numbers) on the screen. The first PC graphics displays were produced by the IBM Colour Graphics Adaptor (CGA). Increasingly powerful resolution is provided by subsequent graphics standards such as Enhanced Graphics Adaptor (EGA), Virtual Graphics Array (VGA) and Super VGA (SVGA).

Monitor

A desktop machine uses a monitor, which is very like a television set. Portable machines have, until now, used a liquid crystal display (LCD) because they require only a fraction of the power needed by the cathode ray tube (CRT) used in the normal desktop monitor. New types of screen are being introduced all the time, and new colour portables are available using special thin-film transistor screens. Nearly all portables with LCD screens have backlighting, which diverts some electrical power to light the screen.

It is important to have a monitor that copes with the anticipated computing needs as well as being easy on the eyes. Desktop machines with higher-resolution monitors, such as 1024 × 768, are now becoming common, with a range of colours that can produce near photographic quality pictures. Monitors vary in size; 14in is the standard but 17in and 21in are now readily obtainable at reasonable prices. Non-interlaced monitors are also becoming more common, as they eliminate the flicker that can be detected on the normal monitors or television screens.

Memory

The computer chip needs memory, in which it stores and runs the programs. All the contents of this random access memory (RAM) are lost when the machine is turned off. For technical reasons, the basic usable memory of an IBM-compatible PC is 640K (1K, or kilobyte = 1024 bytes = 1024 characters). This basic limitation can be overcome by installing extra RAM on separate chips. This is usually installed in multiples of megabytes (1 megabytes = 1000K). It is generally recognized that 4Mb of RAM is needed to run Windows programs successfully and 8Mb is preferable.

Memory that empties when the machine is turned off defeats the object of storing work. Computers have always been expected to store programs and data, and PCs do so by using special magnetic devices known as disks. Machines now come as standard with hard (or Winchester) disk drive units and in addition with a floppy disk drive whose floppy disks can be transferred between machines. The hard disk drive can store much more data than a floppy disk can, and spins at a much greater speed, allowing the data to be accessed much more quickly. The storage capacity of a hard disk drive a few years ago was 20 Mb, and this was thought to be generous. It is now common to find a storage capacity of 100 Mb or even 200 Mb. The good thing is that the price of storage is going down.

Floppy disks come in two sizes, 5.25in and 3.5in, with the latter now more standard. Each size comes in both double-density and high-density formats. A high-density disk drive unit is able to read disks of both low density and high density. Alternatively, tape streamers (magnetic tape similar to that used in cassette recorders) can be used for the storage of computer data.

A compact disk read-only memory (CD-ROM), might also be desirable for the storage of very large programs. A 600 Mb storage capacity is not uncommon. The drive must be correctly interfaced with the computer and special software is needed to get at the information on the disk.

Printers

There is little point in having a computer without a printer. The paperless office is probably an unrealistic ideal. There are essentially three types of printer:

- *Dot matrix printers* have a head made up of a matrix of pins, which are hammered down onto a ribbon to transfer ink to the paper. This is similar principle to an ordinary typewriter. The first dot matrix printers had 9 pins but 24 pins are now standard and the quality of output is reasonably good.

- *Bubble or inkjet printers* have a print head made up of a matrix of tiny nozzles from which beads of ink are squirted out on to the paper. These have advantages of being relatively quiet and producing a good quality of print that rivals some laser printers.

- *Laser printers* fuse powder (toner) on to the paper using a laser. They are quiet, print very quickly, and produce work of high quality.

Software

Standard software can be purchased through mail order. It is heavily discounted in price. As with most things in life you get what you pay for. With computer software you usually get 50% more in features than is necessary or required. The major software packages include the five common types of software; word processing, spreadsheets, databases, communications and desktop publishing. It is also possible to purchase an integrated package of these items. In addition, surveyors are likely to find accounting and graphics (for presentation) software useful. The selection of the particular software depends to some extent upon its familiarity and the perceived future needs. It is most important that the data files are produced in standard format in order that they can be used by the other programs, particularly where the importing or exporting of data is required. It is also important to be careful about the source of any software. There is always the risk of computer virus infections or a prosecution by a manufacturer or FAST (Federation Against Software Theft) where the software has not been legitimately obtained.

Operating systems

An operating system is a set of programs that control the resources and components of the computer. It consists of a number of specialized programs, each of which performs a specific task. One such task, controlled by the operating system, is the allocation of the computer's RAM used by the application programs. The operating system is also responsible for the synchronization of other hardware components such as the monitor, printer and disk drives. Most microcomputers use a disk-based operating system called DOS, and the commands executed within the operating system are known as DOS commands.

An alternative to using the DOS prompts in programs is to choose to work through Windows. This is a product of the Microsoft corporation, which enhances their operating system and makes the computer easier to use. It uses WIMP (windows, icons, mouse and pointers, or pull-down menus). A window can be as big as the screen or as small as you want it to be. Several windows can be open at the same time. This enables the work to be organized on the screen just as it might be spread out in books or papers on a desk.

General functions

These include word processors, which most practices now use. These have the advantages of storage of standard documents for their reuse, a range of word tools such as spell checkers, and thesauruses, and now even grammar checkers. In addition, the quality of printers now allows the inexpensive production of documents of textbook quality. For improved presentations there are also available easy-to-use desktop publishing programs. Spreadsheet programs allow arithmetic and logic operations on numeric data to be performed on the cells of the worksheet and are suitable for a wide variety of tasks. Databases are effectively computerized filing systems, which enable the data to be manipulated and produced in many different ways to suit different purposes. Records can thus be stored for retrieval against different designated criteria.

Communications

Electronic mail is an alternative to the commercial postal services or oral messages sent over the telephone lines. The postal services are relatively slow and telephone messages are sometimes lost. There are systems that can store and deliver text and messages by electronic means. Electronic mail systems include telex, facsimile, communicating text processors, message-switched networks and computer-based message systems. They allow printed material to be passed to others in different geographical locations. They have three main applications:

- as an alternative to mail services
- as a complement to the telephone service
- an optional medium for conferences and meetings.

Some programs include a digital telephone answering machine, fax store, voice mailbox and data modem, all working on one telephone line. Others include desktop processing units, which are able to combine scanning, copying, faxing and printing facilities.

Networking

The purpose of data networks is to interconnect computers and other storage devices that work with computer-coded data, so that the data can be transferred from one location or source to another. Many companies already network their computers but often just to share a printer or to exchange files. But the emphasis is now changing, with the goal of many managers to have a PC on every desk, all connected with employees sharing, communicating and editing the company data. As far as the users of information technology are concerned, data networks operate 'behind the scenes', but they exert a profound influence on the application of IT. For example, access to remote information systems, the operation of electronic mail and the world's increasing telecommunication services are all dependent on the transmission of digital data.

Local area networks (LANs) aim at interconnecting large numbers of different types of data equipment in a single location. They have been developed specifically to deal with data and are faster, cheaper and less prone to errors. Networks allow each microcomputer or workstation to share the high-speed disk storage devices, common information and data and the fast, efficient and high-quality printers.

Quantity surveying practice

Fig. 7.1 provides an indication of the use of information technology in quantity surveying practices. The data is based upon a sample of firms. While the use of IT continues to grow, thanks to its efficiency and effectiveness of operation, in some areas of work manual methods still remain competitive in terms of familiarity and the speed of application. In other cases computerized systems have been able to improve the quality of service provided to clients and to produce information that was not previously available or was too difficult to obtain from manual systems. In some quantity surveying practices IT has progressed to the point where they no longer see the necessity for extensive secretarial support, as their surveyors access most of their information and data directly from the computer. However, typically only about 60% of practices currently have staff who have had some formal computer training.

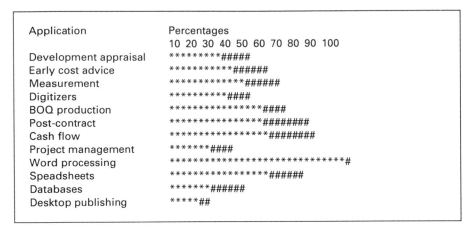

Fig. 7.1   Extent of use of information technology.

***** Current use
Projected use

(*Source:* Based on I. Kirby, 'Making the most of IT', *Chartered Quantity Surveyor*, September 1991.)

Quantity surveying applications

Precontract

Initial price estimating and cost planning
Several different attempts have been made towards using computers for initial price estimating and cost-planning purposes. Some of the emphasis has been placed on seeking to improve the quality of the accuracy of the early price estimates. Others have sought to adapt new estimating techniques, which before the availability of computers would have been impossible tasks owing to their complexity and the time required to complete the process. The Building Cost Information Service (BCIS) has produced its own approximate estimating package, which allows allocation of historic costs to the various elements of the building project. Other systems have sought to include aspects of life-cycle costing and the derivation of the developer's budget.

Both approximate estimates and cost plans can be effectively and easily computed through the use of spreadsheet programs. The advantage of using this approach is that the spreadsheet easily allows the user to quickly revise the cost plan or estimate for changes that may be made to the design of the project.

BCIS on-line
The BCIS is perhaps the largest disseminator of construction cost information in the world. Since 1962 it has been a very successful venture for the acquisition and exchange of cost data amongst its subscribers. Subscribers to the on-line service are able to transfer most of this information directly onto their own computers via a modem attachment to their telephone line. Since new data is constantly being incorporated into the system all the on-line users have access to the latest

information in the database. In addition to this fast access to the data the user is able to manipulate the data to suit particular needs and circumstances.

Estimating databases

Construction cost information of different kinds and in different formats has been in use for over 150 years. The growing complexity of the design and construction of projects and the requirements of clients has generated a need for increasing volumes of cost and price information. The large amounts of data that are now available make computer access to them essential for their effective operation. Many of the industry's estimating databases are now available on a computer disk for easy access, retrieval, application and updating. Often in conjunction with this is software that allows for bills of quantities or work schedules to be rapidly scanned into the system for ease in pricing. This avoids the time-consuming process of rekeying data.

Expert systems (IKBS: intelligent knowledge-based systems)

An expert system is a computer program that has a means of capturing the knowledge of the experts in a particular subject discipline. The computer program, when developed, uses a representation of human expertise to perform functions similar to those performed by the expert. It provides the facility to diagnose problems and produce reasons for any conclusions or recommendations that it makes. Quantity surveying provided a first amongst the construction professions through the development of ELSIE (LC: Lead Consultant). The characteristics of expert systems can be briefly can be briefly identified as follows:

- They know a great deal about a small subject area.
- They are able to provide their advice conversationally.
- Their knowledge is not contained in conventional programs but in separate modules containing sets of rules.
- They operate in areas of activity where uncertainty prevails.
- The questions posed by expert systems are limited to ones that are relevant to a particular line of reasoning.
- The systems are able to explain and justify their reasoning so that human experts can accept their credibility.

Some experts question the ability of a computer to be able to undertake their jobs. Others accept that such a system or something similar will become a natural aid to problem-solving. The professions are concerned that their knowledge, if it can be retained and utilized on a computer, will then become available to a much wider group of people. If the information is to be used effectively it will still require a certain level of expertise to interpret the findings and offer the appropriate advice. It also relies upon the interrogator to ask the correct questions at the outset!

An expert system must contain the following three components:

- a database of expert knowledge;

WOLVERHAMPTON COLLEGE

- a set of rules to manipulate this data which will be more simple than an application program;
- the ability to draw the correct conclusions from the application of this knowledge.

Perhaps the best test of an expert system is to place the computer in another room and simulate a situation where the user is unaware whether the information is being given by a person or machine. To develop the knowledge for the expert system, researchers spend time picking the brains of specialists to extract and structure the knowledge that is a basis for the specialist's expertise.

Contract documentation

Digitized measuring systems
A digitizer is an electronically sensitive drawing board from which dimensions may be 'scaled' directly from drawings into the system. Lengths, perimeters and areas of both simple and complex shapes can be scaled. Incorporated into the digitizer is a facility known as the menu. This is a chart that contains preprogrammed instructions and is used to complement the keyboard, reducing the need for the measurer to remember instruction codes or be a proficient typist. The digitizer uses the normal computer keyboard in conjunction with the measuring process.

The first step is to set up the measurement file to hold descriptions, dimensions and calculations for a particular section of work. The next stage is to attach the drawing to the surface of the digitizer, and to calibrate it according to its dimensions. A standard library of descriptions is used. The measurer then enters a code via the keyboard for the work to be measured, and the dimensions are obtained via the digitizer. Perhaps the most well-known system using this technique is CATO (Computerized Taking-Off). Other systems are available that use light pens for measuring from drawings, as an alternative to the digitizer.

Computer-aided design
A distinction needs to be made between CAD and computer draughting. The latter largely replaces or assists in draughtsmanship, and therefore provides only a minimal input into the design. The CAD system helps designers more directly with their work. Some of these systems have a facility for generating quantities directly and for developing the appropriate specification clauses. The process of measuring could in theory thus be largely removed from the surveyor's responsibility. However, the more likely scenario will be for quantity surveyors to have their own CAD workstations not only for contract documentation production but also for providing the necessary and up-to-date cost information on the different design solutions as the scheme progresses.

Electronic tendering
Electronic data interchange (EDI) is the electronic transfer of business data from one independent computer system to another using agreed standards to format the data. The receiving computer is then able to understand and react to the

information, extracting and processing the data that is needed for the company's use. Such systems are now available that allow, for example, bills of quantities to be issued to contractors in an electronic form, which can then be used by contractors to price the work based upon the estimator's analysis of outputs and other factors that will affect the eventual price charged. The advantages of this are that it speeds up the transmission and receipt of information, reduces the paperwork involved and eliminates the wasteful repetition of rekeying information into a computer system (which is time-consuming and error-prone).

In 1987, the UK Construction Industry Forum set up Electronic Data Interchange in Construction (EDICON). This organization's aim is to ensure that the UK construction industry reaps all the possible benefits of EDI.

Cost management

Integrated management systems

Traditionally, many of the functions undertaken by quantity surveyors have been unrelated to each other. Once the initial price estimate was complete a separate process was then started for the production of the contract documentation. Information produced elsewhere by architects or contractors was beyond the quantity surveyor's boundaries and thus outside their systems.

Initially, when computer systems were first developed they were stand-alone packages. However, in more recent times computer-based systems have been introduced to allow each of the different processes used by surveyors and others to be part of an integrated system of data management. For example, the early price estimate can be refined, as the design develops, into project documents. These in turn can be used for tendering purposes including the allocation of items to a network, which generates contract times and associated costs. RIPAC (from Construction Software Services Partnership), for example, is such a system; it provides a bill of quantities measurement and a post-contract administration system combined.

Integrated databases

The growing use of IT in the construction industry has allowed the speeding-up of the manual process and also the removal of some of the tedious and repetitive aspects of the work. It has also allowed data to be manipulated and analysed.

The term 'database' is used to indicate any large collection of data stored on a computer. In order for different databases to be efficient they need to be connected (integrated) to each other in such a way that the different levels of information can be applied and used for different purposes. While quantity surveyors have their 'own' database, this is really a subset of a larger construction and property database, with this in turn being a part of the worldwide industry and commerce database. The quantity surveyor is possibly the most active data user in the construction process.

The value of databases is that users can find information more easily, more efficiently and more consistently. In the more complex forms the data is expressed in such a way that any changes in a piece of information will have implications of a change in other associated items. The integration of the data in this way provides a common currency of use within a common time-scale of events so that all relevant data is automatically updated.

Other applications

Most quantity surveying activities that were formerly paper-based are now being replicated or replaced by computer applications. These include, for example: life-cycle costing, which relies upon the manipulation of the different variables to test the sensitivity of the various assumptions used; development appraisals and feasibility studies; cost planning; value engineering; and cash flow forecasting. The list of applications is endless, not to mention the multiplicity of software houses and the developed system, some of which have been produced in conjunction with quantity surveying practices and contracting organizations.

Further considerations

The surveyor will also want to consider the following in connection with use of hardware and software:

Security

The use of computers is subject to the provisions of the Data Protection Act 1984, the Copyright, Designs and Patents Act 1988 and subsequent regulations, and the Computer Misuse Act 1990. Obviously, sensitive data should be hidden from unwanted intrusion. Whether this is done through physical means (locking or the removal of drives) or through a password protection is up to the user.

In addition it is important that data being used is protected against possible loss due to system failure. This is normally achieved through the frequent saving of working files and by making back-up copes of data. Many of the larger firms are now recommending that back-up data is stored elsewhere in the event of a loss of data.

Virus control

A computer virus is a program designed to replicate and spread on its own, preferably without the user knowing that it exists. Not all computer viruses cause damage; some are harmless and some are just nuisances. Virus control is achieved either through an anti-virus system, which can be built into the software system, or through programs that can be run to detect whether a virus exists. These controls need to be frequently updated, since they are generally only able to detect known viruses. One popular program claims to be able to detect over 1300 known viruses. Installed software provides for continuous virus scanning and cleaning from infected workstations.

Compression programs

Software and applications have a voracious appetite for disk space. Software is available that can be installed to double the capacity of the hard disk.

Organizers
These are essentially computer-based versions of the ubiquitous personal organizer which can be used for storing and retrieving the information electronically. They are particularly useful where the surveyor has ready access to a computer workstation.

Scanners
These are devices that will convert published text, drawings or photographs into digital information, which a computer can then access or put into storage.

Uninterrupted power supplies (UPS)
These perform two functions: they regulate the mains electricity supply, and provide an independent power source for a short period if the mains supply fails.

Major issues

Machiavelli made this comment in the sixteenth century:

> It should be borne in mind that there is nothing more difficult to handle, more doubtful of success, and more dangerous to carry through than initiating changes. The innovator makes enemies of all these who prospered under the old order, and only lukewarm support is forthcoming from those who would prosper under the new. Their support is lukewarm partly from fear and partly because men are generally incredulous, never really trusting new things unless they have tested them by experience.

It is not surprising therefore that the use of the computer moved slowly in the early days. The removal of a measure of control from people and handing this over to a machine is one of the biggest changes that mankind can make. The 1980s saw these attitudinal changes among surveyors.

Figure 7.2 identifies the top five problems of using computers in practice. Meeting project deadlines remains the major problem area. Some difficulties in the past have occurred where a computer system has failed with the surveyor's data locked inside. Not only does this provide an embarrassment, it frightens surveyors from the possible future use of computers. The reliability of computers

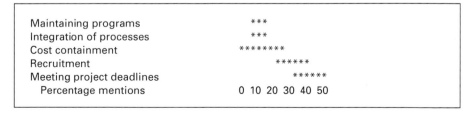

Fig. 7.2   Top five information technology problems.

(*Source:* Price Waterhouse Computing Opinion Surveys.)

has now largely overcome this problem. However, the importance of making back-up copies of work in progress cannot be overstressed. Meeting project deadlines has always been a particular problem for quantity surveyors because of their position of being the last line in the design process and the possible cause of over-runs on the other consultants' activities.

Impact on the quantity surveying profession

IT has its origins in the technologies related to a restricted view of information: the generation, processing and distribution of representations of information. Examples of such technologies are telecommunications, computer engineering, data processing and office machinery. The products of these industries still form the bulk of IT products. Progress in recent years has been towards the extension of data engineering or telematics to an increasing range of areas of application. This has brought with it an active interest in the human aspects of information, such as its quality, value and utilization.

Revolutions of this type, although they appear to take place dramatically quickly, do not happen overnight. A full working lifetime has expired since the advent of the first computer. Also, many of the predictions about the implications of computers have yet to take place, such as the paperless office, or the non-necessity for office space because we shall all be working from home. Even the demise of the bill of quantities has yet to occur. As with most forecasts of the future, individuals and organizations are often wide of the mark. Working life continues to evolve, perhaps gaining in momentum with future change now being the order of the day. The key word in all of the professions is 'adaptability', with the development of transferrable skills rather than skills that will be left behind as being of no future use.

The impact of computers on the quantity surveying profession can be summarized as follows:

- reduction in the amount of time spent on repetitive processes;
- improvements in methods of communications, particularly worldwide;
- enhancement in the quality of the services provided (emphasis is already being placed upon this during this decade of quality);
- development of a broader range of services, sometimes encroaching on other professional disciplines;
- speed in the execution of tasks.

The future

In *Information Horizons* by I. Miles *et al.* three perspectives are provided on the nature of how information technology may change society in the future:

The continuists

Information technology is merely the current stage in a long-term process of developing technological capacities. The so-called revolutionary claims are overstated. The rate of diffusion of IT will be much slower than claimed by interested parties. There will be many mistakes and failures and discouraging experiences. The main features of society are liable to remain unchanged by the use of IT. The future will largely be based upon an extrapolation of the past.

The transformists

IT is based upon revolutionary technology, with unprecedented progress made in computers and telecommunications. The positive demonstration, effects and proven success of IT in meeting new social and economic needs will promote rapid acceptance of change. IT will foster a major shift in industrial and agricultural practices through the introduction of new technologies.

The structuralists

IT has revolutionary implications for the economic and social structure of society. Some countries, industries and professions will become far more adept at capitalizing on the possible potential of IT. Change will appear in waves, demanding new styles, structures and skills. In practice, the future cannot be adequately forecasted; personal development is therefore required that will provide for flexible and adaptable solutions.

Bibliography

Ashworth, G. 'EDI: a world without paper'. *Chartered Quantity Surveyor*, September 1991.
Ashworth, G. 'Electronic tendering moves closer'. *Chartered Quantity Surveyor*, June 1993.
Atkin, B.L. *CAD Techniques: Opportunities for Chartered Quantity Surveyors*. RICS Books, 1987.
Brandon, P.S. (ed.) *Building Cost Modelling and Computers*. Spon, 1987.
Brandon, P.S. 'The development of an expert system for the strategic planning of construction projects'. *Construction Management and Economics*, Spring 1990.
Brandon, P.S. (ed.) *Quantity Surveying Techniques: New Directions*. Blackwell Scientific Publications, 1992.
Brandon, P.S. and Kirkham, J.A. *An Integrated Database for Quantity Surveyors*. The Royal Institution of Chartered Surveyors, 1989.
Brandon, P.S. and McDonagh, N. 'Finding a framework for the future'. *Chartered Surveyor*, April 1991.
Dixon, T. *Computerised Information Systems for Surveyors*. RICS Books, 1988.
Gooch, B. *The Computer Guide for Quantity Surveyors*. B. Gooch, 1984.
Hargitay, S. and Dixon, T. *Software Selection for Surveyors*. Macmillan/RICS Books, 1991.
Institution of Civil Engineers *Applications of Information Technology in Construction*. Thomas Telford, 1991.
Jarrett, M. 'Computerisation: reaping the rewards'. *Chartered Quantity Surveyor*, September 1990.
Kelly, G. 'Computer update'. *Chartered Quantity Surveyor* (various).
Kelly, G. 'Measurement: the end of an era'. *Chartered Quantity Surveyor*, December 1992/January 1993.
Kirby, I. 'Making the most of IT?' *Chartered Quantity Surveyor*, September 1991.
Kirkwood, J. *Information Technology and Land Administration*. Estates Gazette, 1984.

Miles, I. *et al. Information Horizons: The Long-Term Implications of New Information Technologies.* Edward Elgar, 1984.

Wager, D. and Wilson, R. *CAD Systems and the Quantity Surveyor.* Construction Industry Computing Association, 1986.

Chapter 8

Cost control

Introduction

Whenever anyone proposes to construct a building or engineering structure they will need to know in advance the probable costs of construction. These costs include the cost of the works carried out on site by the contractor, professional fees, and any taxes that may be due to the government. In addition to these sums the client will also need to make provision for the costs of the site, other development costs, and the furnishings required in the completed project. These costs are often excluded from the normal process of project cost control. It is one of the duties of the client's quantity surveyors to ensure that the building to be constructed is carefully controlled in terms of costs throughout the entire design and construction process.

Cost control

The term 'cost control' in the construction industry is used to cover the whole service required to meet this end. This process starts at inception, when guide prices will be required, through the stage when an early price estimate is prepared, based upon the client's requirements to the contractors. If the estimate is acceptable and within the client's budget the project then moves to the design stage. In their design the architect or engineer needs to consider alternative solutions to the various aspects of the design. The quantity surveyor will be required to offer comparative costs of the alternative materials to be used or the form of construction to be adopted. This stage, known as cost planning, has been developed in further detail and may be described as a system of relating the design of buildings to their cost so that, while taking full account of quality, utility and appearance, the cost is planned to be within the economic limit of expenditure.

Cost planning further requires the quantity surveyor to divide the approximate estimate into subdivisions, usually of defined elements within the building. These subdivisions are then available for comparison with other cost records. The contract documents may also be prepared on this basis to facilitate easier preparation of the cost analysis.

Cost control does not stop at tender stage but continues up to the agreement of the final account, and the final certificate for the works. Post-contract cost control and procedures are described later in this chapter and in Chapter 11.

Cost advice

The quantity surveyor may be called upon to advise the client on matters of cost at various stages during the design and construction process. Such advice will be necessary regardless of the method used for contractor selection or tendering purposes. However, the advice can be particularly crucial during the early stages of the project inception. During this time major decisions are taken affecting size and quality of the works, if only in outline form. The cost advice given must therefore be as reliable as possible, so that clients can proceed with the greatest amount of confidence.

Quantity surveyors are now recognized within the construction industry as cost consultants. Their skills in measurement and valuation are without equal. It should be recognized, however, that clients and, in their turn, designers who are either unable or unwilling to provide proper information by way of brief, quality or budget, must therefore expect the cost advice to be equally imprecise. The surveyor must also realize the importance of providing realistic cost information that will contribute to the overall success of the project. In this context it is important to be familiar with both design method and construction organization and management.

Quantity surveyors are the industry's experts on building costs and must perform their duties to appropriate professional standards. The client when paying for a professional opinion requires it to be sound and reliable. Failure to carry out these duties properly could provide grounds for liability for negligence. Surveyors should therefore avoid giving estimates 'off the cuff', and it is preferable during the early stages to offer a range of prices rather than just a single figure. There is an old and very true maxim that the first figure is always the figure that the client remembers.

Precontract methods

The following are the methods which can normally be used for pre-tender price estimating. Although they are often referred to as approximate estimating methods, this needs to be read in the context of the way in which they are put together rather than in terms of accuracy of price alone. The degree of accuracy will depend upon the type of information provided to quantity surveyors, the quality of relevant available pricing information and their skills and experience.

Most of the methods are reasonably well known, and are listed below. Only those that are current or have possible future use have been described.

- unit method
- superficial method
- superficial-perimeter method
- cube rules
- storey-enclosure method
- approximate quantities
- cost planning
- cost modelling
- financial methods.

Unit method

The unit method of approximate estimating consists of choosing a standard unit of accommodation and multiplying this by an appropriate cost per unit. The technique is based upon the fact that there is usually a close relationship between the cost of a construction project and the number of functional units that it accommodates. The standard units may for example represent the cost per theatre seat, hospital bed space or car park space. Such estimates can only be very approximate and vary according to the type of construction and standard of finish, but in the very earliest stages offer a guide to a board or building committee.

Superficial method

This type of estimate is simple to calculate and costs are expressed in a way that should be readily understood by the average client. The area of each of the floors is measured and then multiplied by a cost per square metre. In order to provide some measure of comparability between various schemes the floor areas are calculated from the internal dimensions of the building: that is, within the enclosing walls.

The best records for use in any form of approximate estimating are those derived from the surveyor's previous projects and past experience. Extensive use is made of cost databases to store analyse and retrieve such information. Difficulty, of course, occurs where the surveyor has no previous records. In these circumstances the BCIS or the technical press such as *Building* magazine may need to be consulted. The BCIS *Quarterly Review of Building Prices* is also a useful guide to which reference can be made. This type of information must, however, be treated with great caution as it can produce misleading results. It can rarely ever be used without some form of adjustment.

Buildings within the same category, such as schools or offices, have an obvious basic similarity, which should enable costs within each category to be more comparable than for buildings from different categories.

It is also important to consider varying site conditions. A steeply sloping site must make the cost of a building greater than it would be for the same building on a flat site. The nature of the ground conditions, and whether they necessitate expensive foundations or difficult methods of working, must also be considered.

The construction to be used will also have an important influence. A single-storey garage of normal height will merit a different rate per square metre from that of one constructed for double-decker buses. Again, a requirement for, say, a 20 m clear span is a different matter from allowing stanchions at 5 m or 7 m intervals. Further, an overall price per square metre will be affected by the number of storeys. A two-storey building of the same plan area has the same roof and probably much the same foundations and drains as a single storey of that area, but has double the floor area. However, the prices for high buildings are increased by the extra time involved in hoisting materials to the upper floors, and the use of expensive plant such as tower cranes.

The shape of a building on plan also has an important bearing on cost. A little experimental comparison of the length of enclosing walls for different shapes of the same floor area will show that a square plan shape is more economical than a long and narrow rectangle, and that such a rectangle is cheaper than an L-shaped plan. There are, however, exceptions to these general rules.

The standard of finish naturally affects price. There will be clients who require office blocks with the simplest of finish and there will be others to whom, perhaps, more lavish treatment has advertisement value: they may want expensive murals or sculpture.

In projects offering different standards of accommodation it will be preferable to price these independently. A variety of rates may therefore be required, depending upon the different functions or uses of the parts of the building. There may also be the possibility of the need to include items of work that do not relate to the floor area, and these will have to be priced separately.

Approximate quantities

These provide for a more detailed approximate estimate than any of the other methods described. They represent composite items, which are measured very broadly by combining or grouping typical bill measured items. In practice only items of major cost importance are measured. For example, strip foundations are measured per lineal metre and include excavation, concrete and brickwork items up to dpc level. A unit rate for this work can be readily built up on a basis obtained from a cost database price book or priced bills of quantities from previous projects. Walls may include the internal and external finish, and windows and doors can be enumerated as extra over.

This method does provide a more reliable means of approximate estimating, but it also involves more time and effort than the other methods during its preparation. No specific rules exist, but the composite items result from the experience of the individual surveyor. In order for the quantities to be realistically measured more information is required from the designer. Specially ruled estimating paper is available, which is designed particularly for approximate estimating purposes.

Approximate quantities should not be confused with a bill of approximate quantities. The latter would usually be based upon an agreed method of measurement, but since the design is not at a sufficiently advanced stage, the quantities must therefore only be approximate. Also, the approximate quantities used for precontract estimating would be much briefer, as several of the bill items would be grouped together to form a single composite description.

The use of approximate quantities for precontract cost control can create some costing and forecasting difficulties, as by the time the drawings have reached the required stage many of the matters of principle have already been settled. Often these cannot be altered without a major disturbance of the whole scheme.

Cost planning

Cost planning is not simply a method of pre-tender estimating, but seeks also to offer a controlling mechanism during the design stage. Its aim in providing cost advice is to control expenditure and also to offer to the client better value for money. It attempts to keep the designer fully informed of all the cost implications of the design. In today's world full cost planning would also incorporate the attributes of life-cycle costing and value analysis.

Two alternative forms of cost planning have been developed, although in practice a combination of both is generally used. The first type, known as elemental cost planning, was devised by the then Ministry of Education in the early 1950s. It was developed largely in response to the extensive school building programme in which there was found to be a wide variation in cost, and where costs needed to be monitored and controlled more effectively than previously. It is sometimes referred to as 'designing to a cost'. At about the same time the RICS set up its Cost Research Panel and introduced comparative cost planning, which became known as 'costing a design'. The Cost Research Panel was later instrumental in developing the Building Cost Advisory Service (now the Building Cost Information Service).

The cost-planning process commences with the preparation of an approximate estimate and then the setting of cost targets, which are based upon elements. As the design evolves these cost targets are checked for any under- or overspending against the architect's details. The prudent surveyor will also always be looking for ways of simplifying the details, without altering the design, in an attempt to reduce the tender sum. Not only will the building construction be considered but also the ease or otherwise with which the design can actually be built. If the process is carried out satisfactorily it should at least result in fewer abortive designs; it should not be thought to cease at tender stage but to continue throughout the post-contract cost-control procedure. The details of cost planning are beyond the scope of this book and the reader needing more information should refer to the bibliography at the end of this chapter.

To carry out the cost-planning service the surveyor needs the ready and willing co-operation of the architect and client, as the latter will be responsible for the additional fees incurred. Clients should, however, be encouraged to adopt these procedures, and the advantages should be clearly identified to them; indeed most clients now expect that this service will be provided. Public authorities and group practices have an advantage in having both the architect and quantity surveyor mostly in the same building and both subject to a common discipline. There is no reason, however, why the private sector cannot work together in a similar manner if they give their time willingly to achieve the objectives.

Apart from providing a full system of cost planning the surveyor can be of use to the architect by assisting with the comparative costs for alternative systems of construction or finishings. This may involve, for example, comparing different plan shapes in terms of cost, or external cladding. The particular problem must be examined and, for materials, their cost and methods of fixing must be investigated, where these are not always known. The surveyor may also be asked for advice on the total cost implications of a particular method of construction.

Cost limits

These are sometimes referred to as financial methods of precontract cost control. A cost limit is fixed on the building design, based upon either a unit of accommodation or rental value. The architect must then ensure that the design can be constructed within this cost limit. There are common procedures to be adopted in connection with public buildings such as houses, schools and hospitals. They are also used in the private sector where a developer may need to place a limit on the building cost based upon the other costs involved such as the price of the land and the selling price of a house or the commercial value of the project.

Cost modelling

This is a modern technique to be used for forecasting the estimated cost of a proposed construction project. Although cost models were first considered during the early 1970s there is only scant evidence of their use in practice. A statistical model or formula can be constructed which best describes the building in terms of cost. The development of cost model building can be a lengthy process requiring the collection and analysis of large quantities of data. Prior to using the model in practice it needs to be tested against the more conventional methods. Cost modelling is a radical approach to precontract estimating and cost control, and further research is necessary before the techniques can be used with confidence in practice.

Life-cycle costing

It has long been recognized that it is not satisfactory to evaluate the costs of buildings on the basis of their initial costs alone. Some consideration must also be given to the costs-in-use that will be necessary during the lifetime of the building. Life-cycle costing is an obvious idea, in that *all* costs arising from an investment decision are relevant to that decision. The primary use of life-cycle costing in construction is in the evaluation of alternative solutions to specific design problems. The life-cycle cost plan is a combination of initial, maintenance, replacement, energy, cleaning and management costs. Life-cycle costing must take into account the building's life, the life and costs of its components, inflation, interest charges, taxation and any consideration that may have a financial consequence on the design.

Value analysis

Value analysis or value engineering was developed as a specific technique during the 1940s, and has been extensively used for a variety of purposes, particularly in the USA. It has been described as a system for trying to remove unnecessary costs before, during and after construction. It is an organized way of challenging these costs and is based upon a functional analysis that requires the answer to the six basic 'what if' questions of value analysis. In essence, the technique seeks to improve the value for money in construction projects by improving their usefulness at no extra cost, by retaining their utility for less cost, or by combining their improved utility with a decrease in cost.

During the 1950s value engineering started to penetrate European manu-facturing industries and it is today recognized as a means of efficiency-oriented management. It has only in recent years begun to be applied in the UK construction industry. Value analysis is seen to complement current quantity surveying practice and procedure.

Risk analysis

Traditionally, quantity surveyors often presented their clients with single-price estimates, even though it was apparent that on virtually every project differences would occur between this sum and the final account. The construction industry is subject to a greater amount of risk and uncertainty than most other industries. Also, unlike many other major capital items, construction projects are not developed from prototypes; each scheme represents a bespoke solution, often involving untried aspects of design and construction, in order to satisfy an individual client's requirements. All construction projects include aspects of risk and uncertainty.

Risk is measurable and can therefore be accounted for within an estimate. For example, quantity surveyors involved in forecasting the costs of a new project will have access to different sorts of cost data and this, coupled with their expertise, will enable a budget price range to be calculated within specified confidence limits. It is desirable to offer an estimate in this way rather than to suggest a single price. Uncertainty is more difficult to assess, as it represents unknown events that cannot be even assessed or costed. Different techniques can be applied, such as Monte Carlo simulation, to assess the risk involved. The risk will of course not be eliminated but at least it can be managed rather than ignored.

Computerized systems

Most of the above processes are now undertaken using computer systems in one form or another. This may involve computerizing an existing method of working in order to remove repetition and calculation when preparing pre-tender price estimates. Alternatively, it may result in the adoption of specialized software that has been designed to ease and improve the processes involved. Such software may be coupled with cost databases or may utilize the developing expert systems that are now becoming available. These are referred to further in Chapter 7.

General considerations

The selection of appropriate rates for pre-tender estimating and cost control depends upon a large variety of factors. Some of these can be considered objectively, but in other circumstances, only experience or 'feel' for the project may suffice. Some of these factors are discussed below:

Market conditions

The surveyor, in order to get the price as near as possible to the tender sum, must be able to interpret trends in prices based upon past data and current circumstances. This demands a great deal of skill and some measure of luck! Allowances may also need to be made for changes in contractual conditions, type of client, labour availability, workloads, and the general state of the industry.

Design economics
When using previous costs or cost analyses, changes in these costs for design variables such as shape, height and size will need to be taken into account.

Quality factors
Cost information is deemed to be based upon defined or assumed standards of quality. Where the quality in the proposed project is considered to be different from this, then changes in the proposed estimate rates will be required. The surveyor should always provide quality indications with any cost advice.

Engineering services
These are becoming an ever-increasing portion of building projects. Their cost importance is such that they now need to be considered separately. On large schemes it is now usual to employ specialist quantity surveyors who are experienced in matters of building services.

External works
Owing to the considerable differences that often exist between building sites, there are few established cost relationships for this element of buildings. The size of the site and the nature of the work to be carried out will be important factors to consider.

Exclusions
The proposed estimate of cost should clearly identify what has been included, by way of the outline specification and, equally important, what has been excluded. Clients may well assume that a tender sum includes all of their costs and will be concerned, to say the least, should they find that certain items were excluded. Obvious items include fees, VAT, land costs, loose furniture and fittings. In some projects the whole of the fitting-out work may form part of a separate contract.

Price and design risk
If at any stage during the design and construction process the design is still being evolved, then a contingency sum will need to be provided for possible additional design costs. This is design risk. The contingency sum is therefore likely to be larger at the early stages of planning than at the tender stage. The price risk factor is largely related to market conditions. A market that is more volatile than usual will result in a larger percentage being allowed in the estimate for this factor.

Accuracy of approximate estimates

Whether the process of cost planning at the design stage is used or not, an analysis of tenders will be valuable for future approximate estimating purposes. Quantity surveyors must attempt to anticipate matters that will affect the level of future tender sums. They cannot, however, forecast exceptional swings in existing tendencies, just as they cannot forecast the future. They must, however, be aware of all current developments, which they will often find referred to in the journal of their institution and other technical information, such as *Building* and BICS. Any serious qualification of an estimate should be promptly reported, and not left to contribute to an explanation of an unexpectedly high tender.

The accuracy of the estimate is of prime importance. On average it would appear that ±10% is the typical sort of figure expected in the construction industry. Forecasts are unlikely to be consistently spot-on, as by definition estimates will be subject to some degree of error. It is preferable to present estimates as a range of values rather than as a single lump sum. Alternatively, confidence limits could be offered as a measure of estimates' reliability. The following factors are said to have some influence upon the accuracy of estimating:

- the quality of the design information;
- the amount, type, quality and accessibility of cost data;
- the type of project, as some schemes are easier to estimate than others;
- the project size, as accuracy increases marginally with size;
- the stability of market conditions;
- the familiarity with a particular type of project or client.

Proficiency in estimating is the combination of many factors, including skill, experience, judgement, knowledge, intuition and personality.

Preparing the approximate estimate

The method to be used for the preparation of an approximate estimate will to some extent depend upon the type of project, together with the amount of information that the client is able to provide: The more vague the design information, the less precise will be the estimate. The estimate for a complicated refurbishment project could be prepared on the basis of the gross internal floor area, but this would require a large quantity of other calculations on the part of the quantity surveyor. Of course, if the surveyor was familiar with the type of work, the designer and the client, then this approach would be satisfactory.

A better method for such a type of project would be to use some form of approximate quantities. Usually, once the estimate was acceptable, then this would form a budget for the project. It may be desirable to prepare the estimate in accordance with the RICS publication *Pre-contract Cost Control and Cost Planning*.

It is first of all necessary to quantify the project using one of the methods described earlier and then to price these quantities using cost data. The cost data may be obtained from previous projects (that is, from priced bills), from cost analyses, or from some published source of cost data. It is usual to add on a contingency amount to cover unforeseen items of work. The amount of this is highest at inception and it gradually reduces as the design becomes more firm. It is never entirely removed until the completion of the final account.

As the project is priced at current prices some addition is required to allow for possible increased costs. This is normally added as two separate amounts: the first up to tender stage in order to allow for comparison with tenders, and a second sum to allow for increased costs during construction. The forecasting of these sums in periods of high inflation is at times hazardous.

Approximate estimates are normally given exclusive of VAT, even on those projects where VAT will be charged. This can represent a considerable item so this should be made quite clear to the client. Even this total sum will not represent the full costs of construction to the client. It is also important to be aware if a particular scheme is to be exempt from VAT: for example, the client is entitled to charitable status (see also Chapter 9). Professional fees for the architect, engineer and quantity surveyor, and other charges for planning approval, must also be added. The professional fees will always attract VAT unless the practice is a very small one.

The estimate should also be clear as to items that have been excluded altogether. Generally the distinction between client's and contractor's items will be straightforward but where confusion might occur it is better to spell this out in the budget estimate. A typical cost plan summary is shown in Fig. 8.1.

Post-contract methods

A variety of methods are used to control the costs of construction during the post-contract stage of development. If they are to be effective, all changes to the contract sum must be costed prior to instructions for variations being issued.

Budgetary control

Budgets are used for planning and controlling the income and expenditure. It is through the budget that a company's plans and objectives can be converted into quantitative and monetary terms. Without these a company has little control. A budget for a construction project represents the contract sum divided between a number of different sub-headings or work packages. The contractor will have a costed work programme for the project, although this can be disrupted through changes (variations) to the scheme or the acceleration or the deceleration of activities. The client's budget represents the time-scale of payments and the availability of funds for honouring the contractor's certificates. Clients with several projects under construction will need to aggregate the amounts of interim certificates from different projects to obtain the total funding requirements. In addition, clients are concerned with the total forecasted project expenditure. The ability to control this depends upon the sufficiency of the precontract design, the need for subsequent variations, the steps taken to avoid unforeseen circumstances, and matters which are beyond their control, such as strikes.

The contractor's budget will provide a rate of expenditure and a rate of income throughout the project. The contractor's funding requirements represent the difference between these two items, and the amount of capital required at the different times can then be calculated. Contractors also need to aggregate this information from all their current projects in order to determine the company position. Budgetary control compares the budgets with the actual sums incurred, explaining the variances that arise. In common with other control techniques,

JOB Nr: 123   Date: January 1994

New Church of England Primary School
Blankchester Diocesan Board of Finance

COST PLAN
(Fixed Price 65 weeks)   *Floor area* 1200 m²

Summary	Cost per m² £	Elemental cost £	£
1. SUBSTRUCTURE	54.00	64 800	64 800
2. SUPERSTRUCTURE			
2.1 Frame	32.67	39 200	
2.2 Roof	43.08	51 700	
2.3 External walls	20.08	24 100	
2.4 Windows and external doors	20.75	24 900	
2.5 Internal walls and partitions	15.67	18 800	
2.6 Internal doors	11.17	13 400	172 100
	143.42		
3. INTERNAL FINISHINGS			
3.1 Wall finishings	15.17	18 200	
3.2 Floor finishings	19.75	23 700	
3.3 Ceiling finishings	11.00	13 200	55 100
	45.92		
4. FITTINGS AND FURNISHINGS	21.25	25 500	25 500
5. SERVICE INSTALLATIONS			
5.1 Sanitary appliances	8.33	10 000	
5.2 Disposal installation	11.67	14 000	
5.3 Hot and cold water installation	20.83	25 000	
5.4 Heating installation	52.50	63 000	
5.5 Electrical installation	26.67	32 000	
5.6 Gas installation	2.50	3 000	
5.7 Communication installation	3.33	4 000	
5.8 Builders work in connection	7.50	9 000	160 000
	133.33		
6. EXTERNAL WORKS			
6.1 Site work	104.16	125 000	
6.2 Drainage	19.75	23 700	
6.3 External services	8.33	10 000	158 700
	132.24		
7. PRELIMINARIES	54.17	65 000	65 000
8. CONTINGENCIES			35 000
TOTAL ESTIMATED COST	£584.33 per m²		£736 200

Exclusions: Professional fees and site supervision
          Building control fees
          Site investigation costs
          Abnormal ground conditions
          Value added tax

Fig. 8.1 Sample cost plan summary.

budgetary control is a continuous process undertaken throughout the contract duration.

Clients' financial reports

Financial reports are prepared at frequent intervals throughout the contract period, depending upon the size and complexity of the project, to advise clients on any expected changes to the contract sum. An example of a typical financial statement will be found in Chapter 11, where such reporting is covered in more detail.

Clients' cash flow

In addition to the client's prime concern with the total project costs, the timing of cash flows is also important. The client's advisors will prepare an expenditure cash flow based upon the contractor's programme of activities. On large and complex projects and in periods of high inflation the timing of payments, based upon different constructional techniques and methods, might result in higher tender sums representing a better economic choice for the project as a whole.

Contractor's cost control

The contractor, having priced the project successfully enough to win the contract through tendering, must now ensure that the work can be completed for the estimated costs. One of the duties of the contractor's quantity surveyor is to monitor the expenditure, and advise management of action that should be taken. This process also includes the cost of subcontractors, as these are likely to form a significant part of the main contractor's total expenditure. The contractor's surveyor will also comment on the profitability of different site operations. Wherever a site instruction suggests a different construction process from that originally envisaged then details of the costs of the site operations are recorded. The contractor's surveyor will also advise on the cost implications of the alternative construction methods that might be employed.

Discounting the fact that estimators can sometimes be wide of the mark when estimating, even with common work items, the contractor would seek reasons to justify a wide variation between costs and prices. This will be done for two reasons: first, in an attempt to recoup, where possible, some of the loss; and secondly to remedy such estimating or procedural errors in any future work. There are various reasons why such variations may arise:

- The character of the work is different from that envisaged at the time of tender.
- The conditions for executing the work have changed.
- Adverse weather conditions severely disrupted the work.
- There was an inefficient use of resources.
- There was an excessive wastage of materials.
- Plant had to stand idle for long periods of time.
- Plant had been incorrectly selected.

● Delays had occurred because of a lack of accurate design information.

This list is not exhaustive, and often when the project is disturbed by the client or designer this can also have a knock-on effect on the efficiency and outputs of the contractor's resources generally. Contractors may also suggest that they always work to a high level of efficiency. This is not always the case, and the loss is sometimes due to their own inefficiency. Costing that shows that a project has lost money is of limited use where the contractor cannot not remedy the situation. The contractor needs to be able to ascertain which part of the job is in deficit and to know as soon as it starts to lose money. The objectives therefore of a contractor's cost control system are:

● to carry out the works so that the planned profits are achieved;
● to provide feedback for use in future estimating;
● to cost each stage or building operation, with information being available in sufficient time so that possible corrective action can be taken;
● to achieve the benefits suggested within a reasonable level of administration charges.

Contractor's cash flow

Contractors are not, as is sometimes supposed, singularly concerned with profit or turnover. Other factors also need to be considered in assessing the worth of a company or the viability of a new project. Shareholders, for example, are primarily concerned with the rate of their return on the capital invested. Contractors have become more acutely aware of the need to maintain a flow of cash through the company. Cash is important for day-to-day existence, and some contractors have suffered a downfall not because their work was unprofitable but because of an insufficiency of cash in the short term. In periods of high inflation, poor cash flows have resulted in reduced profits, which in their turn have produced an adverse effect for the shareholders' return. It is necessary therefore to strike the correct balance between the objectives of cash flow, profit, return and turnover. In addition, inflation and interest charges will also have an influence upon these items.

Bibliography

Aqua Group, The *Pre-contract Practice for the Building Team*. Blackwell Scientific Publications, 1992.

Ashworth, A. *Cost Models, Their History, Development and Appraisal*. CIOB Technical Information Service No. 64, Chartered Institute of Building, 1986.

Ashworth, A. *Cost Studies of Buildings*. Longman, 1994.

Ashworth, A. and Skitmore, M. 'Accuracy in cost estimating'. Proceedings of the Ninth International Cost Engineering Congress, Oslo, 1986.

Association of Cost Engineers *Estimating Checklist for Capital Projects*. Spon, 1991.

Barrett, F.R. *Financial Reporting, Profit and Provisions*. CIOB Technical Information Service No. 12, Chartered Institute of Building, 1982.

BCIS *Quarterly Review of Building Prices*. BCIS (various).

Brandon, P.S. (ed.) *Building Cost Modelling and Computers*. Spon, 1987.

Brandon, P.S. (ed.) *Quantity Surveying Techniques: New Directions*. Blackwell Scientific Publications, 1992.

Bull, J. (ed.) *Life Cycle Costing for Construction*. Blackie, 1993.

Burgess, R.A. *Construction Projects and their Financial Policy and Control*. Longman, 1980.

Cooke, B. and Jepson, W. *Cost and Financial Control in Construction*. Macmillan, 1982.

Fellows, R.F. 'Cash flow and building contracts'. *The Quantity Surveyor*, September 1982.

Ferry, D.J. and Brandon, P.S. *Cost Planning of Buildings*. Blackwell Scientific Publications, 1990.

Flanagan, R. and Norman, G. *Life-Cycle Costing for Construction*. RICS Books, 1983.

Kelly, J. and Male, S. 'Value management: enhancing value or cutting costs?' RICS Occasional Paper, The Royal Institution of Chartered Surveyors, 1991.

MacPherson, J., Kelly, J. and Male, S. *The Briefing Process: A Review and Critique*. The Royal Institution of Chartered Surveyors, 1992.

Nicholson-Cole, D. 'Cash flow information for the client'. *Architects Journal*, October 1983.

Nisbet, J. 'Post-contract control, a sadly neglected skill'. *Chartered Quantity Surveyor*, January 1979.

Raftery, J. *Models for Construction and Price Forecasting*. The Royal Institution of Chartered Surveyors, 1992.

Raftery, J. *Principles of Building Economics*. Blackwell Scientific Publications, 1991.

RICS *Precontract Cost Control and Cost Planning*. The Royal Institution of Chartered Surveyors, 1982.

Somerville, D.R. *Cash Flow and Financial Management Control*. CIOB Surveying Information Service No. 4, Chartered Institute of Building, 1981.

Trimble, E.G. and Kerr, D. 'How much profit goes to the bank?' *Construction News*, March 1974.

Chapter 9

Procurement

Introduction

Procurement is the process used to obtain construction projects. The dictionary definition states that procurement is 'acquiring or obtaining by care or effort'. It involves the selection of a contractual framework that clearly identifies the responsibilities of the different parties involved.

A little over 30 years ago the clients of the construction industry had only a limited choice of procurement methods available to them for commissioning a new construction project. Since then there have been several catalysts for change in the procurement of construction projects, such as:

- Government intervention;
- pressure groups being formed to create change for the benefit of their own members, most notably the British Property Federation;
- international comparisons, particularly with the USA and Japan, and the influence of the Single European Market;
- the apparent failure of the construction industry and its associated professions to satisfy the perceived needs of its customers in the way that the work is organized;
- the influence of developments in education and training;
- the impact from research studies of contracting methods;
- the response from industry, especially in times of recession, towards greater efficiency and profitability;
- changes in technology, particularly information technology and computing;
- the attitudes towards change and improved procedures from the professions;
- the clients' desire for single-point responsibility.

General matters

The selection of appropriate contractual arrangements for any but the simplest type of project is difficult because of the diverse range of views and opinions that are available. Much of the advice is conflicting and lacks a sound base for evaluation. Individual experiences, prejudices, vested interests and familiarity, together with the need for change and the real desire for improved systems, have all helped to reshape procurement options towards the end of the twentieth century. The proliferation of differing procurement arrangements has resulted in an increasing demand for systematic methods of selecting the most appropriate arrangements to suit the particular needs of clients and their project. The following are the broad issues involved.

Consultants or contractors

These issues relate to whether to appoint independent consultants for design and management or to appoint a contractor direct. The following should be considered:

- single-point responsibility
- integration of design and construction
- need for independent advice
- overall costs of design and construction
- quality, standards and time implications.

Competition or negotiation

There are a variety of different ways in which designers or constructors can secure work or commissions, such as invitation, recommendation, speculation or reputation. However, irrespective of the final contractual arrangements that are selected, the firms involved need to be appointed. Evidence generally favours some form of competition in order to secure the most advantageous arrangements for the client. There are, however, many different circumstances that might favour negotiation with a single firm or organization. Some of these are listed below; however, it must not be assumed that the choice between opting for competition and negotiation is clearly defined, as each case must be decided on its own merits:

- business relationship
- early start on site
- continuation contract
- state of the construction market
- contractor specialization
- financial arrangements
- geographical area.

Measurement or reimbursement

There are in essence only two ways of calculating the costs of construction work. The contractor is either paid for the work executed on some form of agreed quantities and rates or reimbursed the actual costs of construction. The following are the points to be considered between the alternatives:

- necessity for a contract sum
- forecast of final cost
- incentive for efficiency
- distribution of price risk
- administration time and costs.

Traditional or alternative methods

Traditionally, most projects built in this century in the UK have used single-stage selective tendering as their basis for contracting. With a wider knowledge of the different practices and procedures around the world, and some dissatisfaction with this uniform approach, other methods have evolved to meet changing circumstances and aspirations by clients. The following factors should be considered; these are examined in more detail later in this chapter:

- appropriateness of the service
- length of time from inception to completion
- overall costs inclusive of design
- accountability
- importance of design, function and aesthetics
- quality assurance
- organization and responsibility
- project complexity
- risk apportionment.

Standard forms of contract

There are a wide variety of different forms of contract in use in the construction industry. The choice of a particular form is dependent upon a number of different circumstances, such as:

- private client or public authority
- type of work to be undertaken
- status of the designer
- size of proposed project
- method used for price determination.

Local authorities use different forms of contract from central government departments, while nationalized industries and some of the larger manufacturing companies have developed their own forms and conditions. These often place a greater risk on the contractor and this is turn reflected in the contractor's tender prices. The different industry interests continue to develop a plethora of different forms for their own particular sectors. A trend towards greater standardization would be welcomed, and the introduction of new forms resulting in even greater fragmentation is to be deprecated. As long ago as 1964, the Banwell Report recommended the use of a single form for the whole of the construction industry as being both desirable and practicable. This message has largely gone unheeded owing to the variety of interested parties involved.

Although the general layout and contents of the various forms are similar, their details and interpretation may vary immensely. Different forms exist for main and subcontracts, for building or civil engineering works, and according to the relationship between the client, consultants and contractors. The different versions of the JCT (Joint Contracts Tribunal) forms of contract

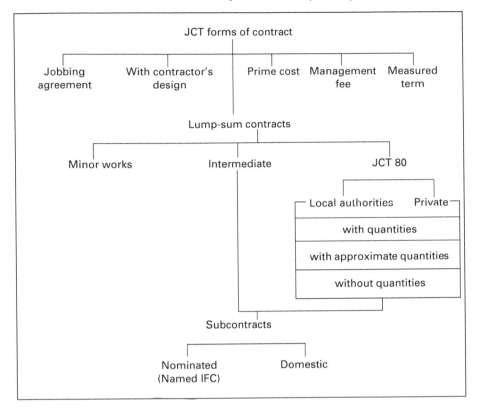

Fig. 9.1   JCT forms of contract.

Source: Adapted from J. Parris, *The Standard Form of Building Contract*, Blackwell Scientific Publications.

(Fig. 9.1) are used on most building contracts, with JCT 80 the most frequently used form on major projects. On civil engineering projects the Conditions of Contract and Form of Tender, Agreement and Bond for use in connection with works of Civil Engineering Construction remains the popular form of contract, although the Institution of Civil Engineers have recently launched the New Engineering Contract (NEC), which they claim will be appropriate for all types of construction work.

In addition there are the General Conditions of Government Contracts for Building and Civil Engineering Works (GC/Works/1 for Lump Sum Contracts with/without quantities, Single Stage Design and Build and GC/Works/2 for Minor Works Lump Sum and Prime Cost Contracts) published by HMSO, forms produced by the Association of Consultant Architects (ACA), the British Property Federation (BPF), the Conditions of Contract (International) for Works of Civil Engineering Construction (FIDIC), and forms published by the Institution of Chemical Engineers (I Chem E) and the National Economic Development Council (NEDC) for process plants and fabrication for the oil and gas industry.

Practice notes

The Joint Contracts Tribunal from time to time issue 'practice notes', which express their view on some particular point in practice. While due account should be taken of such opinions, they do not affect the legal interpretation of the terms of the contract and are thus not finally authoritative. They are similar to a discussion in Parliament of the interpretation of an Act.

Methods of price determination

Building and civil engineering contractors are paid for the work that they carry out on the basis of one of two methods.

- *Measurement*: The work is measured in place (that is, in its finished quantities) and paid for on the basis of quantity multiplied by rate. Measurement may be undertaken by the client's quantity surveyor, in which case an accurate and detailed contract document can be prepared. Alternatively it may be undertaken by the contractor's surveyor or estimator, in which case it will be detailed enough only to satisfy the contractor concerned. Measurement contracts generally allow the predicted final cost to be calculated, and this has clear advantages to the client for budgeting and cost control.

- *Cost reimbursement*: The contractor is paid the actual costs based on the quantities of materials purchased and the time spent on the work by operatives, plus an agreed amount to cover profit. Dayworks are valued on the same basis.

Measurement contracts

The alternative forms of measurement contract used in the construction industry are as follows.

Drawing and specification

This is the simplest type of measurement contract and is really only suitable for small or simple project work. There has, however, in recent years been a trend towards using this method on large projects, often based upon misconceived ideas. Each contractor measures his own quantities from the drawings and specification and prices them in order to determine the tender sum. The method is thus wasteful of the contractor's estimating resources, and does not really allow for a fair comparison of tender sums. The contractor also has to accept a greater risk, since in addition to being responsible for the pricing they are also responsible for the measurements. In order to compensate for possible errors, contractors will tend to overprice the work.

Performance specification

This method results in an even vaguer approach to tendering. The contractor is required to provide a price based upon the client's brief and user requirements alone. The contractor must therefore choose a method of construction and type of materials suitable for carrying out the works. The contractor is likely to select the least expensive materials and methods of construction that comply with the laid-down performance standards.

Schedule of rates

In some projects it is not possible to predetermine the nature and full extent of the works. In these circumstances a schedule is provided that is similar to a bill of quantities, but without any quantities. Contractors then insert rates against these items, and these will be used to calculate the price based upon remeasurement. This procedure has the disadvantage of being unable to provide a contract sum, or any indication of the likely final cost of the project. On other occasions a comprehensive schedule already priced with typical rates is used as a basis for agreement. The contractor in these circumstances adds or deducts a percentage adjustment to all the rates. This standard adjustment can be unsatisfactory for the contractor, as some of the listed rates may be high prices and others low prices. The best known schedule of this kind is the PSA Schedule of Rates.

Bill of quantities

This is still the most common type of document used for a measurement contract. It provides the best basis for estimating tender comparisons and contract administration. The contractors' tenders are therefore judged on price alone as they are all using the same measurement data. This type of documentation is usually recommended for all but the smallest projects.

Bill of approximate quantities

In some instances it may not be possible to measure the work accurately. In this case a bill of approximate quantities would be prepared and the entire project remeasured upon completion.

Cost-reimbursement contracts

These types of contract are not favoured by many of the industry's clients as there is an absence of a tender sum and a possible final account cost. This type of contract also often provides little incentive for the contractor to control costs. They are therefore only used in special circumstances, for example:

- emergency work projects, where time cannot be allowed for the traditional process;
- when the character and scope of the works cannot be readily determined;
- where new technology is being used;
- where a special relationship exists between the client and the contractor.

Cost-reimbursement contracts can take several different forms. The following are three of the types that may be used in the above circumstances. Each of the methods pays contractors' costs and makes an addition to cover profit. Prior to embarking upon this type of contract it is important that all the parties concerned are fully aware of the definition of contractors' costs as used in this context.

Cost plus percentage

The contractor is paid the costs of labour, materials, plant, subcontractors and overheads, and to this sum is added a percentage to cover profits. The percentage is agreed at the outset of the project. A disadvantage of this method is that the contractor's profit is related directly to expenditure. Therefore the more time that is spent on the works the greater will be the profitability. In practice, because it is an easy method to operate, this tends to be the method selected when using cost reimbursement.

Cost plus fixed fee

In this method the contractor's profit is predetermined by agreeing a fee for the work before the commencement of the project. There is therefore an incentive for the contractor to attempt to control the costs, because it will increase the rate of return. In practice, because it is difficult to predict the cost accurately beforehand, it can cause disagreement between the contractor and the client's professional advisers when trying to settle the final account, if the actual cost is much higher than that which was estimated at the start of the project.

Cost plus variable fee

The use of this method requires a target fee to be set for the project prior to the signing of the contract. The contractor's fee is made up of two parts: a fixed amount, and a variable amount depending upon the actual cost. This method provides an even greater incentive to the contractor to control costs, but has the disadvantage of requiring the target cost to be fixed on the basis of a very rough estimate.

Contractor selection

There are essentially two ways of selecting a contractor: through competition or by negotiation. Competition may be restricted to a few selected firms or open to almost any firm that wishes to submit a tender. The options described later are used in conjunction with one of these methods of contractor selection. Contractors are sometimes requested to provide a monetary deposit for the documents, which is then returned to them upon the submission of their tender. This is an attempt to discourage the long tender lists of firms that have been recorded on some projects. There have been recorded instances where firms have had to pay non-returnable fees to join the list of tenderers: a practice that has been severely criticized by the industry watchdogs.

A Code of Tendering Procedure has been developed by the National Joint Consultative Committee (NJCC), which although not mandatory does provide guidance and good practice on the awarding of construction contracts. The Code is frequently reviewed to take into account the changes in building procurement practices. The points covered in the Code include use of a standard form, limitations on the numbers of tenderers, preliminary enquiry information, the submission of tenders on the same basis, time available for tendering, withdrawal of tenders, and the method to be used for the correction of any errors.

Selective competition

This is the traditional and most popular method of awarding construction contracts. In essence a number of firms of known reputation are selected by the project team, sometimes following a form of prequalification, to submit a price for the project based upon the contract documents. The firm that submits the lowest tender is then awarded the contract. The 'known reputation' of the firm includes:

- the firm's financial standing and record;
- its recent experience of building over similar contract periods;
- the general experience and reputation of the firm for similar building types;
- the adequacy of its management;
- its capacity to undertake the project.

Open competition

With this now rarely used method of contract procurement the details of the proposed project are first of all advertised in local or trade publications. Contractors who feel able to carry out the proposed work request copies of the tender documents. This allows new contractors or those who are unknown to the project team and the client the possibility of submitting a tender for consideration. In theory any number of firms are therefore able to submit a price. However the costs of tendering can be high and the process of pricing the items lengthy, and therefore there should be a limit on the number of tenderers. Also, such costs must be borne by the construction industry and will be absorbed in successful tenders. The use of open tendering may relieve the client of the obligation of accepting the lowest price. This is because firms are generally not vetted before the tenders are submitted. Factors other than price must therefore be taken into account when assessing tender bids. It is generally accepted that a lowest-price tender will be obtained by the use of this method. There is, of course, no obligation on the part of the client to accept any tender, should none of the contractors or their prices be considered suitable.

Negotiated contract

This method of contractor selection involves the agreement of a tender sum with a single contractor. The contractor will offer a price using the tender documentation, and this is then reviewed in detail by the client's quantity surveyor. The two parties then discuss the rates that are in contention, and through a negotiation process a tender acceptable to both parties can be agreed. Owing to the absence of any competition or other restriction other than the acceptability of price, this type of contract procurement is not generally considered to be cost-advantageous. It usually results in a tender sum that is higher than might have been obtained by using one of the previous methods.

A negotiated contract should result in fewer errors in pricing. It may involve the contractor in some participation during the design stage, and this may result in savings in both time and money. It should also be possible to achieve greater cooperation during the construction period between the designer and the contractor. Because of the higher sums incurred, public accountability and the suggestion of possible favouritism, local government does not generally favour this method.

Contractual options

The following contractual options are an attempt to address the client's objectives associated with time, cost and quality of construction. They are not mutually exclusive. For example, it is possible to award a serial contract using in the first instance a design-and-build arrangement. Fast tracking may also be used in association with the other options that are available. In addition to these various options, it will need to be decided whether the contractor selection is to be through competition or negotiation.

Traditional

In this approach the client commissions an architect to take a brief, produce designs and construction information, invite tenders and administer the project during the construction period and settle the final account. If the building owner is other than small, the architect will advise the client to appoint consultants such as quantity surveyors, structural engineers and building services engineers. The contractor, who has no design responsibility, will normally be selected by competitive tender or there may be good reasons for negotiating a tender. The design team are independent advisers to the client and the contractor is only responsible for executing the works in accordance with the contract documents. Fig. 9.2 shows the relationships of the parties.

Early selection

This method is sometimes known as two-stage tendering. Its main aim is to involve the chosen contractor on the project as early as possible. It therefore tends to succeed in getting the person who knows what to build (the architect) in touch with the firm that knows how to build it (the contractor) before the design is finalized. The contractor's expertise in construction methods can thus be used in the architect's design. A further advantage is that the selected contractor will also be able to start on site sooner than would be the case with the other methods of contract procurement.

In the first instance an appropriate contractor must be selected, usually on the basis of some form of competition. This can be achieved by inviting suitably selected firms to price the major items of work from the project. A simplified bill of quantities is therefore required that will include the preliminary items, major items and specialist items, allowing the main contractor the opportunity of pricing for profit and attendance sums.

The contractor will also be required to state their overhead and profit percentages. The prices of these items will then form the basis for subsequent price agreement that will be achieved through negotiation.

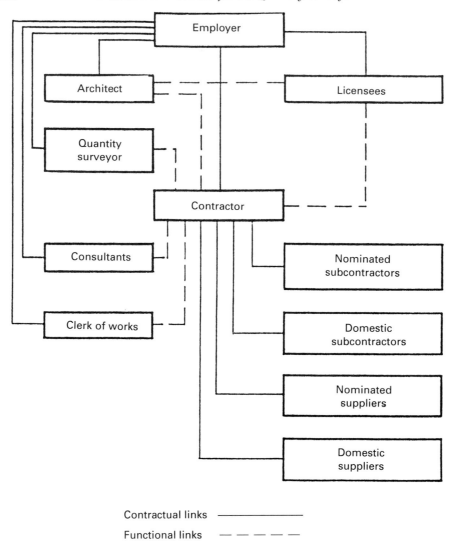

Contractual links   ———————

Functional links   — — — — —

Fig. 9.2   Traditional contract.

Design and build

It has been suggested that the separation of the design and construction processes, which is traditional in the UK construction industry, has been responsible
for a number of problems. Design and build (Fig. 9.3) can often overcome these
by providing for these two separate functions within a single organization. This
single firm is generally the contractor. The client, therefore, instead of
approaching an architect for a design service, chooses to go directly to the
contractor. The client may choose to retain the services of an architect or quantity

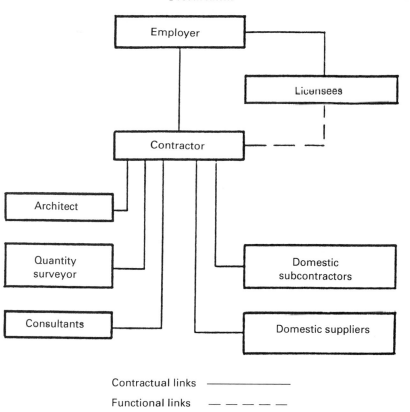

Contractual links —————————
Functional links — — — — —

Fig. 9.3 Design and build.

surveyor to assess the building contractor's design or to monitor the work on site.

The design evolved by the contractor is more likely to be suited to their own organization and construction method and this should result in a saving both in time and cost of construction. It may be argued, of course, that the design is more in keeping with the contractor's construction ability than the particular design needs of the client. The final building should, however, result in lower production costs on site, a shorter design and construction period and an overall saving in price to the client, after taking into account the savings on design fees. A further advantage to the client is the implied warranty of suitability because the contractor has provided the design.

A major disadvantage, which needs to be weighed against the advantages described above, is the discouragement of possible variations by the client. Where clients consider these to be necessary they often have to pay an excessive sum of money for their incorporation within the finished building.

Design-and-build projects usually result in the client's obtaining a single scheme from the selected contractor. Where it is desirable to obtain some form of competition in price, then the type and quality of design will also need to be taken into account. In practice this can present difficulties in the evaluation of the various schemes considered.

Package deal

In practice the terms 'package deal' and 'design and build', when applied to construction projects, are interchangeable. Strictly speaking, however, a package deal is a special type of design-and-build project in which the client chooses a suitable building from a catalogue. The client will also probably be able to view similar buildings that have been constructed elsewhere, of a similar design and type of construction.

This type of contract procurement has been used extensively for the closed systems of industrialized buildings of timber or concrete such as multi-storey office blocks and flats, low-rise housing, industrial premises and farm buildings. The client typically provides the package deal contractor with a site and supplies specific user requirements.

On occasion an architect may be independently employed to advise on the proposed building type selected or to supervise the works during construction. This may be particularly appropriate to those items that are outside the scope of the system superstructure.

The type of building selected is an 'off-the-peg' type structure that can be erected very quickly. However, there is even less scope for variations than with the design-and-build approach, should the client wish to change certain aspects of the construction detailing. As far as costs are concerned, opinions vary. It cannot be automatically assumed that the package deal will always provide a more economic solution to the client's needs, either initially or in the long term.

Turnkey method

This form of contracting is unusual in the UK and has not been used to any large extent. It has, however, been used successfully in both the Middle East and Far East.

The true turnkey contract includes everything from the inception design up to hand-over and the possible furnishings of the project by a single contractor. In some projects it may also include the provision of a suitable site, prior to design and construction. An all-embracing contract is therefore formed with a single administrative entity for the entire building's procurement process. The promoter therefore expects to be able to walk in and take over the project which is then ready for use.

The method gets its name from the 'turning-the-key' concept. It is therefore an extension of the traditional design-and-build contract, and in some cases it may even include the long-term maintenance work. On industrial building projects the appointed contractor may also be required to design and install the manu-facturing equipment. This type of contract planning may therefore be appro-priate for use with highly specialized types of industrial and commercial construction projects.

The entire building procurement and maintenance needs to be handled by a single company. It has been argued, however, that the client's ability to control costs, quality, performance, aesthetics and constructional details is somewhat limited by the nature of the procurement method.

Management contracting

The term 'management contracting' is used to describe a method of organizing the project team and operating the construction process. The management contractor acts in a professional capacity, providing the management expertise and buildability required in return for a fee to cover overheads and profit. The contractor does not therefore participate in the profitability of the construction work itself and does not employ any of the labour or plant, except for the possibility of the work involved in the setting up of the site and the costs normally associated with the preliminary works.

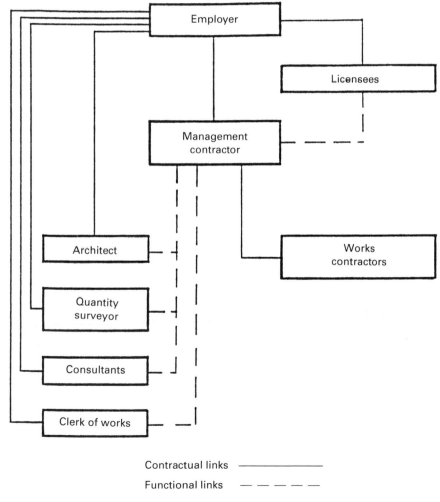

Fig. 9.4   Management contracting.

Because the contractor is employed on a fee basis, the appointment can take place early during the design stage. The contractor is therefore able to provide a substantial input into the practical aspects of the building technology process. Each trade required for the project is tendered for independently by sub-contractors, upon the basis of either measured quantities or a lump sum. This should therefore result in the lowest cost for each trade and for the construction works as a whole. The management contractor assumes the full responsibility for the control of the work on site (Fig. 9.4).

Construction management

This procurement route offers an alternative to management contracting and has been adopted on a number of large projects over recent years. The main difference is that the individual trade contractors are in a direct contract with the client.

The client appoints a construction manager (either a consultant or contractor) with the relevant experience and management expertise. The construction manager, if appointed first, would take the responsibility for appointing the design team, who would usually also be in a direct contract with the client.

The construction manager is responsible for the overall control of the design team and trade contractors throughout both the design and construction stages of the project.

Project management

Although the definition of the term 'project management' is not universally agreed, the following description generally conforms to what is understood in the UK. Project management requires the client to appoint a professional adviser to this post, who will then in turn appoint the appropriate design consultants and select the contractor to carry out the work. The method is appropriate for large buildings and engineering projects. The function of the project manager is therefore to organize and coordinate the design and construction programme. The construction work may then be carried out using one of the other methods that has been described. The role of the quantity surveyor as project management is fully described in Chapter 13.

Management fee contract

Management fee contracting is a system whereby a contractor agrees to carry out building works at cost. In addition the contractor is paid a fee by the client. Some contractors are prepared to offer an incentive on the basis of target cost. This type of procurement is a similar approach to cost-plus contracts and therefore has similar advantages and disadvantages.

Design and manage

This method of contracting (Fig. 9.5) is really the counterpart to design and build. In this case the design manager (architect, engineer or surveyor) has full control, not just of the design stage but also of the construction phase. They effectively

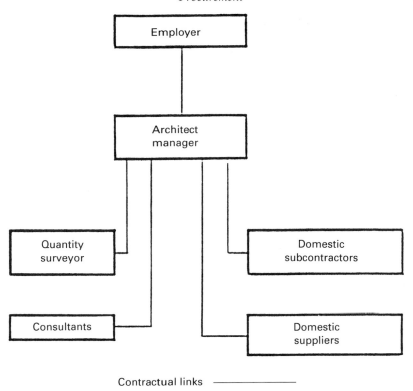

Contractual links ————————

Fig. 9.5 Design and manage.

replace the main contractor in this role, which these days mainly concerns management, administration and coordination of subcontractors. The design manager is responsible for all the aspects of construction including the programming and rectification of any defects. The contract is between the design-and-manage firm and the client, and this offers the client a single line of responsibility.

This method of contractor selection appears to offer all the advantages of traditional tendering coupled with those of design and build. The design-and-manage firm will of course need to engage its own construction managers and as such will need to consider some continuity in this type of work.

It is suitable for all sizes of project but clients undertaking large projects may, because of past experience, prefer to use a traditional form of procurement involving one of the larger contractors. A further disadvantage may lie in the facilities that need to be provided; for the design-and-manage contractor these will always have to be hired.

This type of procurement should be able to offer competitive completion times compared with the other methods that are available. As there is independent control of the subcontractor firms this should ensure quality at least as good as that offered by other contracting methods. As for cost, this should be as competitive as any of the other systems.

The British Property Federation

The British Property Federation (BPF) represents a substantial commercial property interest in the UK, and is thus able to exert some influence on the construction industry, its professions, contractors and the procedures employed. The BPF system unashamedly makes the interests of the client paramount. Its aim has been to devise a more efficient and cooperative method of organizing the whole of the building process from inception up to completion, by attempting to make a genuine use of all the parties involved with design and construction. It was developed largely in response to the dissatisfaction of clients with the existing arrangements; it was claimed that buildings cost too much, took too long to construct (compared with buildings elsewhere in the world) and often did not achieve the standards of quality that were expected.

The BPF manual sets out the operation of the system in detail, and while this may appear to be a rigid set of rules and procedures its originators claim that it can be used flexibly, often in conjunction with other methods of procurement. As the system has almost been entirely devised by one party to the construction contract, it lacks the compromises inherent in the contracts agreed jointly by clients and contractors. Innovations not generally found in the traditional methods include: single-point responsibility for the client, utilization of the contractors' buildability skills, reductions in the pre-tender period, redefinition of risks and the preference for specifications rather than measured quantities.

Fast tracking

This approach to contracting results in the letting and administration of multiple construction for the same project. It is applicable generally to large projects. The process results in the overlapping of the various design and construction operations of a single project. These various stages may therefore result in the creation of separate contracts or a series of phased starts and completions. When the design for complete sections of works, such as foundations, is completed the work is then let to a contractor, who will start on site while the remainder of the work is still being designed. The contractor for this stage or phase will then see this section through to completion. This staggered letting of the work has the objective of shortening the construction time for the overall project, from inception to hand-over to the client.

This type of contract will require considerably more organization and planning on the part of the architect. In practice, where this method is required a project manager will usually be appointed. Although the hand-over date of the project to the client will be much earlier, this may be at the expense of the other factors of cost and quality.

Measured term

This type of contract is often used for major maintenance projects. It can be awarded to cover a number of different buildings. It will usually apply for a specific period of time, although this may be extended, depending upon the necessity of maintenance standards and the acceptability of the contractor's performance. The contractor will at the outset be offered the maintenance work for various trades. The work when completed will then be paid for using rates from an agreed schedule. This schedule may have been prepared specifically for the project concerned, or it may be a standard document such as the PSA Schedule of Rates.

Where the client supplies the rates for the work, the contractor is given the opportunity of quoting a percentage addition to or deduction from these rates. The contractor offering the client the most advantageous percentage will be awarded the contract. An indication of the amount of work involved over a defined period of time would therefore seem appropriate for the contractor's assessment of the prices quoted. The JCT publishes a Measured Term Form of Contract suitable for use with any schedule of rates.

Serial tender

Serial tendering is a development of the system of negotiating further contracts, where a firm has successfully completed a contract for work of a similar type. Initially contractors would tender against each other, possibly on a selective basis, for a single project. There is, however, a legal understanding that several other similar projects would automatically be awarded using the same bill of rates. Some allowances would be made to cover inflation. The contractors would therefore know at the initial tender stage that they could expect to receive a number of contracts, which would provide them with continuity in their workload. Conditions would, however, be written into the documents to allow further contracts to be withheld where the contractor's performance was unsatisfactory. Serial contracting should result in lower prices to contractors since they are able to gear themselves up to such work by, for example, purchasing suitable types of plant.

Serial contracts are appropriate to buildings such as housing and schools in the public sector. This method may also be usefully employed in the private sector in the construction of industrial units. It has been successfully used with industrialized systems buildings.

Firm-price contracts

Provisions exist under most forms of contract for the alternative arrangements of either a fixed price or a fluctuating price agreement. The choice usually depends upon the length of the contract period and the amount of inflation present in the economy. Where the inflation factor is in single figures and falling, it is usual to award fluctuating contracts for a project's duration exceeding two years. For contracts on a firm-price basis, therefore, the contractor must estimate the possible differences due to fluctuations in costs and add these to the tender prior to submitting it to the architect, quantity surveyor or client. It cannot automatically be assumed that a firm-price arrangement will necessarily be more beneficial for the client.

Value added tax and building contracts

Value added tax (VAT) was introduced to the construction industry through the Finance Act 1972. During the Chancellor's annual Budget statement there is the opportunity to amend both the extent and percentage rate of this tax, and since 1972 this has been done many times. The current legislation is covered in the HM Customs and Excise leaflet 708/2/90 dated 1 August 1990. In addition to this leaflet other provisions cover protected buildings (VAT leaflet 708/1), property development (VAT leaflet 742A), property ownership (VAT leaflet 742B), aids for handicapped persons (VAT leaflet 701/7) and VAT refunds for DIY builders (Notice 719).

Building work is either standard-rated work, (currently 17.5%) or zero-rated work. Examples of zero-rated work include children's homes, old people's homes, homes for rehabilitation purposes, hospices, student living accommodation, armed forces living accommodation, religious community dwellings and other accommodation used for residential purposes. Certain buildings intended for use by registered charities may also be zero-rated. Buildings that are specifically excluded from zero-rating include hospitals, hotels, inns and similar establishments.

The conversion, reconstruction, alteration or enlargement of any existing building is always standard-rated. All services that are merely incidental to the construction of a qualifying building are standard-rated. These include architects', surveyors' and other consultants' fees and much of the temporary work associated with a project. Items that may be typically described as 'furnishings and fittings', such as fitted furniture, domestic appliances, carpets, or free-standing equipment, are always standard-rated irrespective of whether the project may be classified as zero-rated.

The *VAT Guide* (Notice 700) gives examples, but throughout the document individuals are advised to check their respective liability with the local VAT office. The ratings of some items are arbitrary, and some will need to be tested by the courts.

The contract sum referred to in article 2 and clause 14 of JCT 80 is exclusive of value added tax. Adjustments to the contract sum will also be exclusive of the tax, as the conditions specifically state that VAT will not be dealt with under the terms of the contract. If, after the date of tender, the goods and services become exempt from VAT, then the employer must pay the contractor an amount equal to the loss of the contractor's input tax. This will then equate with the amount that the contractor would otherwise have recovered.

Contract strategy

The choice of the correct method to be used for contractor selection and price determination (Fig. 9.6) is a difficult one because of varying opinions amongst construction experts as to their advantages and disadvantages. These opinions are sometimes based only on limited experience and hearsay from other surveyors and members of the construction team. Quantity surveyors are perhaps in the best position to be able to advise on the best course of action for contractor procurement because of their special contractual skills. There is, however, a need

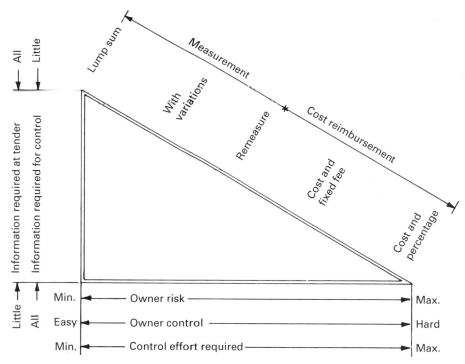

Fig. 9.6 Range of contract types.

Source: Adapted from R.A. Burgess (ed.) *Construction Projects, Their Financial Policy and Control*. Longman, 1980.

to identify more precisely those factors that will allow the profession to offer more consistent and accurate advice in this respect.

However, whichever method is chosen, it should be the one that best serves the client's interest. Clients may at times need to be convinced that particular methods and procedures should be adopted that may be against their own better judgement or prejudice. The quantity surveyor's advice should be offered in a professional manner without any vested interest or personal gain in mind. Clients, particularly those who are unaccustomed to building, rely heavily upon sound professional advice in these and other matters, and this should not be given to suit the surveyor's needs, practice or workload (Fig. 9.7). The following points should be borne in mind.

Size

Small schemes are not suitable for the more elaborate forms of contractual arrangement. Such procedures are also unlikely to be cost-effective. Small schemes will more likely use either selective or open tendering or a type of design and build. Medium to large projects are able to use the whole range of methods; with the very large schemes more advanced and complex forms of procurement may be necessary.

Criterion	Question	Option	Traditional selective tendering	Early selection	Design and build	Construction management	Management fee	Design and manage
Timing	– Is early completion important to the success of the project?	Yes:		*	*	*	*	*
		Average:		*	*	*	*	*
		No:	*					
Variations	– Are variations to the contract important?	Yes:	*	*		*	*	*
		No:			*			
Complexity	– Is the building technically complex or highly serviced?	Yes:	*	*	*	*	*	*
		Average:		*	*	*	*	*
		No:						
Quality	– What level of quality is required?	High:	*	*	*	*	*	
		Average:	*	*	*	*	*	*
		Basic:						
Price certainty	– Is a firm price necessary before the contracts are signed?	Yes:	*		*	*		*
		No:		*			*	
Responsibility	– Do you wish to deal only with one firm?	Yes:	*	*	*		*	*
		No:				*		
Professional	– Do you require direct professional consultant involvement?	Yes:	*	*		*	*	*
		No:			*			
Risk avoidance	– Do you want someone to take the risk from you?	Yes:	*	*	*			
		Shared:				*	*	
		No:						*

Fig. 9.7 Identifying the client's priorities.

Source: Adapted from Thinking Aloud About Building, *EDC report.*

Client

Clients who regularly carry out construction work are much better informed, develop their own preferences and will not require the same level of advice as those who only build occasionally. They may, however, need to be encouraged to adopt more suitable and appropriate methods, and the quantity surveyor will need to convince such clients that the adoption of such suggestions will lead to improved procedures. Local authorities seem to favour selective competition, sometimes against their best interest, whereas some commercial companies believe that a form of management contracting is more relevant to their specific needs.

Cost

It is generally believed that open tendering will gain the lowest price from a contractor. Negotiated tendering supposedly adds about 5% to the contract sum. Projects with unusually short contract periods tend to incur some form of cost penalty. The introduction of conditions that favour the client, or the imposition of higher standards of workmanship than normal, will also push up costs.

Clients, it would appear, are more concerned with achieving the lowest final cost and with having a prediction of cost as accurate and as early as possible. Of less importance are cash flow projections and the indeterminate life-cycle cost. Some have attempted to argue that a directly employed labour force is preferable for some types of contract (when other economies are also taken into account), or that design and build or cost reimbursement is the best for the client. It is very difficult to make realistic and fair comparisons, even where similar projects are being constructed, under different contractual arrangements.

Different methods of contractor selection have been devised in an attempt to overcome this problem. These different approaches must, however, be read in conjunction with the other characteristics that they involve. An optimum time period should be suggested to the client, taking into account these other factors. A superstore may need to be opened 'next' week, whereas a new school will not be needed until the beginning of term. Speculative housing units need to be completed only at a rate at which they can be sold. Considerations include the shortest overall time from inception to hand-over, the shortest period of time the contractor is actually on site (because this affects costs), reliable guaranteed completion dates, and phased completions to fit in with a client's programme.

Design

It can be argued that the best design will be obtained from someone who is independent of the contractor. However, this said, many designs in recent years have not fulfilled the needs of the client, have resulted in considerable defects in use and have lacked 'buildability'. These facts, together with some good marketing on the part of contractors, have encouraged clients to favour design and build. However, even with this evidence, the design of buildings and structures should be undertaken by those who have the relevant expertise and who are independent of the commercial aspects of the construction.

Quality

The statement 'you get what you are prepared to pay for' is certainly true in the context of construction work. Quality in construction is reflected in the design and specification, supervision and capabilities of the contractor. Fast-track construction may impair the quality of the project and the lack of independent supervision may be detrimental. The selection of a firm of contractors who are reputable for the type of work envisaged will go a long way towards satisfying this criterion. The presence of labour-only and other subcontractors, although providing a specialist input to the project, may result in a deterioration in quality control through organizational difficulties.

Organization

Using a separate design does introduce a further tier of responsibility and thereby increases the possibility of things getting overlooked. The traditional methods of contract procurement have set the lines of demarcation quite clearly. The employment of a single firm, as in design and build, does allow the contractor greater freedom to organize the work. Complex, difficult-to-understand arrangements can have the effect of removing from the contractor the much-needed initiative. Management contracting can allow the contractor greater freedom to solve problems as they arise in order to maintain progress.

Market

The selection of the process to be used will vary with the general state of the economy, and an appropriate solution today may not be the correct one tomorrow. When there is ample work available, contractors may be reluctant to enter into what they consider to be unsatisfactory relationships. When prices are low, a form of cost reimbursement may prove to be an expensive proposition. When there is a large amount of work available, workmanship supervision will need to be enhanced.

Risk

There is risk for everyone involved in a construction project. This risk is not entirely associated with money. It can be suggested that some risk can be reduced or even eliminated. The remaining risk can then be shared equally between the contractor and the client. The client's view may, however, be to transfer the major part of this risk to the contractor. Some evidence might suggest that this is not an appropriate course of action to follow. It certainly is not a fair approach.

Role of the quantity surveyor

It is of fundamental importance to clients who wish to undertake construction work that the appropriate advice is provided on the method of procurement to be used. The advice must be relevant and reliable and should offer the best practice using the skills and expertise that are available. It is sometimes difficult to elicit the necessary information from the client and the evidence for a suitable recommendation may be confusing and conflicting. The advice provided is sometimes based upon self-interest, experience and familiarity, with too little reliance upon the evaluation of past performance or future analysis.

Quantity surveyors are in an excellent position as procurement managers with their specialist knowledge of construction costs and contractual procedures. They are able to appraise the characteristics of the competing methods that might be appropriate and to match these with the particular needs and aspirations of the employer. Procurement management may be broadly defined to include the following:

- determining the employer's requirements in terms of time, cost and quality;
- assessing the viability of the project and providing advice in respect of funding and taxation advantages;
- recommending an organizational structure for the development of the project as a whole;
- advising on the appointment of the various consultants and contractors in the knowledge of the information provided by the employer;
- managing the information and coordinating the work of the different parties;
- selecting the methods for the appointment of consultants and contractors.

Procurement procedures are dynamic activities that are evolving to meet the changing needs of society, the industry and its clients. There are no longer standard solutions; each individual project needs to be separately evaluated upon its own individual set of characteristics. A wide variety of different factors need to be taken into account before any sound advice or implementation can be provided. The various influences, at the time of development, need to be weighed carefully and always with the best and long-term interest of the client in mind.

Bibliography

Allen, D. 'Towards the client's objective', in Brandon, P.S. and Powell, J.A. (eds) *Quality and Profit in Building Design*. Spon, 1984.

Aqua Group, The *Tenders and Contracts for Building*. Blackwell Scientific Publications, 1990.

Ashworth, A. *Contractual Procedures in the Construction Industry*. Longman, 1991.

Brandon, P.S. *Intelligent Authoring of Construction Contracts*. The Royal Institution of Chartered Surveyors, 1992.

'Building without conflict'. *Building*, November 1991.

Chappell, D. *Which Form of Contract*. Architectural Press, 1991.

Chappell, D. and Powell-Smith, V. *The JCT Design and Build Form*. Blackwell Scientific Publications, 1993.

CIRIA *Client's Guide to Traditional Contract Building*. Construction Industry Research and Information Association, 1984.

CIRIA *Management Contracting*. Construction Industry Research and Information Association, 1983.

Clamp, H. *The Shorter Forms of Building Contract*. Blackwell Scientific Publications, 1992.

'Contracts in use'. *Chartered Quantity Surveyor*, January 1993.

Cornes, D.L. *Design Liability in the Construction Industry.* Blackwell Scientific Publications, 1994.

Davis, N. 'Package deal services and fees'. *Chartered Quantity Surveyor,* October 1981.

Fish, R. 'Tendering in a competitive manner'. *Chartered Quantity Surveyor,* August 1985.

Franks, J. *Building Procurement Systems.* Chartered Institute of Building, 1990.

Hayes, R. 'The risks of management contracting'. *Chartered Quantity Surveyor,* November, 1985.

Hibberd, P.R. *Subcontracts Under the JCT Intermediate Form.* Blackwell Scientific Publications, 1987.

Hobson, D. *Management Contracting: A Step in the Right Direction.* Surveyors Publications, 1985.

Hutchinson, K. and Putt, T. *The Use of Design/Build Procurement Methods by Housing Associations.* The Royal Institution of Chartered Surveyors, 1992.

MacPherson, J., Kelly, J. and Male, S. *The Briefing Process: A Review and Critique.* The Royal Institution of Chartered Surveyors, 1992.

Morledge, R. 'The effective choice of building procurement'. *Chartered Quantity Surveyor,* July 1987.

NEDO *Faster Building for Commerce.* HMSO, 1988.

NEDO *How Flexible is Construction?* HMSO, 1990.

NJCC *Code of Procedure for Selective Tendering for Design and Build.* National Joint Consultative Committee, 1985.

NJCC *Code of Procedure for Single Stage Selective Tendering.* National Joint Consultative Committee, 1989.

NJCC *Code of Procedure for the Letting and Management of Domestic Sub-Contract Works.* National Joint Consultative Committee, 1989.

NJCC *Code of Procedure for the Selection of a Management Contractor and Works Contractors.* National Joint Consultative Committee, 1991.

NJCC *Code of Procedure for Two Stage Selective Tendering.* National Joint Consultative Committee, 1983.

Parris, J. *Construction Law Digest.* Blackwell Scientific Publications.

Pearson, G. 'Tender assessment of cost plus fixed fee contracts'. *Chartered Quantity Surveyor,* November 1985.

Perry, G. and Thompson, P.A. *Target and Cost Reimbursable Contracts.* Construction Industry Research and Information Association, 1982.

Sharp, P. 'Management or fee or both'. *Chartered Quantity Surveyor,* May 1982.

Sidwell, A.C. 'An evaluation of management contracting'. *Construction Management and Economics,* 1983.

Skitmore, R.M. and Marsden, D.E. 'Which procurement system? Towards a universal procurement selection technique'. *Construction Management and Economics,* Winter 1988.

Turner, A. *Building Procurement.* Macmillan, 1990.

Willis, C.J. and Willis, J.A. *Specification Writing for Architects and Surveyors.* Blackwell Scientific Publications, 1991.

Chapter 10

Preparation of contract documentation

Introduction

Traditionally contract documents comprise:

- form of contract
- specification and/or
- bills of quantities
- drawings.

The various forms of contract have been described in the previous chapter. Once the choice of form of contract has been decided, the next step is the preparation of the documents that will accompany the signed form of contract.

This chapter covers the coordination of the contract documents, and the preparation of a bill of quantities, the main document with which a quantity surveyor has traditionally been concerned. Reference is also made to the preparation of a specification and the obtaining of tenders. The chapter ends with a description of the steps to be taken in collating the formal contract documents themselves.

Coordinated project information

One of the prime causes of disruption of building operations on site has been highlighted as shortcomings in drawn information, together with a lack of compatibility in project information generally: that is, the drawings, specifications and bills of quantities all say something different.

In order to improve the situation, the Coordinating Committee for Project Information (CCPI) was set up by the major bodies in the construction industry. After consultation with all interested parties they produced a Common Arrangement of Work Sections for Building Works (CAWS). They also produced Codes of Project Specification Writing and Production Drawings as well as working with the producers of the Standard Method of Measurement for Building Works (SMM7). CCPI as a working committee has been disbanded but its work is now monitored by the Building Project Information Committee (BPIC).

The purpose of CAWS is to define an efficient and generally acceptable identical arrangement for specification and bills of quantities.

The main advantages are:

- *Easier distribution of information, particularly in the dissemination of information to subcontractors.* One of the prime objects in structuring the sections was to ensure that the requirements of the subcontractors should not only be recognized but should be kept together in relatively small tight packages.

- *More effective reading together of documents.* Use of CAWS coding allows the specification to be directly linked to the bill of quantities descriptions, cutting down the descriptions in the latter while still giving all the information contained within the former.

- *Greater consistency achieved by implementation of the above advantages.* The site agent and clerk of works should be confident that when they compare the drawings with the bill of quantities they will no longer ask the question 'Which is right?'

CAWS is a system based on the concept of work sections. To avoid boundary problems between similar or related work sections, CAWS gives, for each section, a list of what is included and what is excluded, stating the appropriate sections where the excluded item can be found.

CAWS has an hierarchical arrangement in three levels. For instance:

- Level 1 R   Disposal Systems
- Level 2 R1   Drainage
- Level 3 R10   Rainwater pipes/gutters.

There are 24 level 1 group headings, 150 work sections for building fabric and 120 work sections for services. Although very much dependent on size and complexity, no single project will need more than a fraction of this number: perhaps as a very general average 25–30%. Only level 1 and level 3 are normally used in specifications and bills of quantities. Level 2 indicates the structure, and helps with the management of the notation. New work sections can be inserted quite simply without the need for extensive renumbering.

Bills of quantities

Appointment of quantity surveyor

The appointment of the quantity surveyor is likely to have been made at an early stage when early price estimates were under consideration, but when the architect has formalized the working drawings the appointment will have to be settled or, for small contracts, a decision will have to be made as to whether a quantity surveyor is required at all.

A figure of £100 000 gives architects and surveyors a good guide to a reasonable limit of contract size, beyond which in normal circumstances contractors should not be asked to tender in competition without bills of quantities. There may, of course, be special considerations that would justify such invitations for a larger contract and, equally well, similar considerations that could call for a bill of quantities being provided for a contract of a lesser value.

The work involved if six or eight contractors are each asked to prepare their own quantities for estimating, when only one can be successful (and even then the job might be abandoned), is very heavy and obviously uneconomical. Certain public authorities are required by their standing orders to advertise publicly for tenders.

Under the EU regulations public sector construction contracts and private sector contracts financed by more than 50% by public authorities over 1 million currency units (ECUs) (approximately £3 500 000) must be invited and awarded in accordance with the procedures laid down in EEC Directive 89/440. The Directive provides for a 'restricted tendering procedure', which permits the selection of technically and financially competent contractors following advertisement in the *Official Journal of the European Communities* circulating throughout member states.

In such cases there might be five, ten or more contractors tendering, and the sum total of increased overheads if they had to prepare their own quantities would be impossible, to say nothing of the scarcity of qualified staff to do it. The different interpretations which each contractor may put on the same drawings and specification, if there is cause for doubt, are a further disadvantage and a possible cause of differences at a later stage.

Purpose of bills of quantities

The main purpose of a bill of quantities is for tendering. Each contractor tendering for the project is able to price the work on exactly the same information with a minimum of effort. This therefore avoids duplication in quantifying the work, and allows for the fairest type of competition.

Despite the predicted demise of the bill of quantities over 50% of the value of all building work in the UK is still let using lump-sum contracts with firm or approximate quantities. Most other procurement routes such as design and build and management contracting also involve quantification of building work in some form or other. Bills of quantities may not be appropriate for all types of construction work, and other suitable methods of contract procurement are available. For example, for minor works a drawing and specification may be adequate, or where the extent of the work is unknown, payment may be made by using one of the methods of cost reimbursement (see Chapter 9).

In addition to tendering, bills of quantities also have the following uses, and these should be borne in mind during their preparation:

- valuations for interim certificates;
- valuation of variations;
- ordering of materials if used with caution and awareness of possible errors and future variations;
- cost analysis for use in future approximate estimating;
- planning and progressing by the contractor's site planner;
- final accounting;
- quality by reference to the trade preamble clauses;
- domestic subcontractor quotations;
- cost information.

Taking-over of drawings, etc.

The drawings may be sent to the quantity surveyor with specification notes or a complete specification, or there may be an invitation to go to the architect's office to collect them and discuss the job. On such a visit notes should be made of any verbal instructions given, but it is unwise to ask too many questions until some examination has been made of the documents. The questions in mind may be answered by detailed examination of the drawings or perusal of the specification.

The surveyor should ascertain whether it is proposed to supply any further drawings. It may happen that drawings of some section of the work are not complete, and a good deal of time may be wasted if the taking-off of that section is begun on the drawings available. A 1:20 detail to follow may quite likely alter the 1:100 drawing, and an alteration, however slight, may affect a lot of items in the dimensions. If further drawings are to follow, it is helpful if the order in which they are to come can be agreed, having regard both to the architect's office procedure and the surveyor's requirements. If there will be many, a programme with dates should be drawn up to enable the surveyor to organize the work.

Study of documents

The drawings received should be stamped with the surveyor's name and date of receipt except, of course, originals that have to be returned. On a job of any size, a register of drawings should be completed, giving their reference number, scale and brief particulars (see Fig. 10.1). The advantage of separate sheets is that each

DRAWING REGISTER									
Project title: Southtown School Consultant: Architect			Project Nr: 00 Sheet Nr: 1						
Drawing identification			Revisions						
Nr.	Title	Scale							
ss/ 1	Block plan	1:1000							
3	General plan	1:100							
4	Sections, elevations	1:100							
5	Classroom plans	1:20							
6	Cloaks, toilets plans	1:20							
7	Assembly hall plan	1:20							
8	Heating chamber	1:20							
11	Door details	FS							
12	Metal window schedule	1:50							
15	Drainage	1:200							
17	Classroom store fittings	1:20							

Fig. 10.1   Sample drawing register.

taker-off can have a copy, which will assist quick reference until thoroughly acquainted with the drawings. Particulars of any further drawings received should be added to the list when they come in.

The documents will then be examined by a principal or senior assistant in charge of the job and the takers-off to be employed. The first things to be done are: to see that all the necessary figured dimensions are given, both on plans and sections; to see that the figured dimensions are checked with overall dimensions given; and to insert any dimensions that can be calculated and may be useful in the measurement. Any errors in figured dimensions, which are easily made in a final tracing, should be confirmed with the architect, so that the originals can be amended accordingly.

Except on the smallest jobs the drawings should be supplied in duplicate, and on larger jobs there may be three or more copies. It is advisable to number the sets, so that it can be seen at a glance to which set a drawing belongs. This can easily be done by numbering each set 1, 2 or 3 in coloured pencil in the bottom right-hand corner. If these markings, as well as details such as the surveyor's name and the date, are kept in this corner they will be easily visible when drawings are turned up.

It will make the rooms stand out clearly if the walls are coloured in on plan and section in the surveyor's office; moreover, the act of colouring them will give an early indication of the general construction. This can be done quickly with coloured pencil and will be found well worth while. There may also be manuscript notes, or even alternations in plan, made by the architect at the last moment on one copy, and these should be transferred to the other copies. It sometimes happens that plans and sections are only hatched as a labour-saving device (as hatching will be continually printed on all copies) but the surveyor should superimpose his colouring.

It will also be found useful in a job of any size to mark on the general plan (usually 1:100) the positions of the parts that are detailed. A cross-reference in coloured ink in both cases will stand out. There may, for instance, be a number of 1:20 details spread over several sheets: for example, entrance doors, bay windows, or particular points of construction. Sections can be referenced by normal section lines superimposed on the plan in a distinctive colour with the drawing number and selection reference given thus:

meaning that a detail on this line will be found on drawing No. 24, section B-B. Elevations can be referenced in a similar way.

The extent of detail plans can be marked in the same colour with the number of the drawing to be referred to and arrows showing the approximate extent coloured, thus:

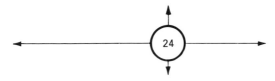

It may be found frequently that sections on detail sheets are not given a letter reference by the architect, but have a title such as 'Section through Kitchen'. The surveyor can give them a letter reference for the special purpose. The marking of the general plans in this way makes them serve as a key, so that a taker-off working on some particular part of the building can see at a glance what details are available.

A careful perusal should then be made of the specification or specification notes. By following through systematically in one's mind the sections of the taking-off one may find gaps in the specification which need filling. These may be quite numerous when only notes are supplied, as 'notes' vary considerably in quality, thoroughness and extent. When a standard specification, such as NBS in one of its versions, is used then the opportunity can be taken to check that the correct alternatives have been chosen, that superfluous matter has been deleted, and that all the gaps have been filled in.

Schedules

Schedules are useful, both for quick reference by the taker-off and for eventual incorporation in the specification for the information of the clerk of works and site agent. They may be supplied by the architect with the drawings, or it may be necessary for the surveyor to draft them. Internal finishings should certainly be scheduled in a tabulated form, so that the finishes of each room for ceiling, wall and floors can be seen at a glance, with particulars of any skirtings, dadoes or other special features. Schedules for windows and doors would include frames, architraves and ironmongery; those for manholes would give a clear size, invert, thickness of walls, type of cover and any other suitable particulars.

Some of the material on schedules may be otherwise shown on drawings, but the schedule brings the parts together and gives a bird's-eye view of the whole. The schedule of finishings, if not supplied in the form of a drawing, should be copied and given to each taker-off, as each will at some time want to know what finish comes in a particular place, when it is necessary to take for deductions or making good.

Queries

As a result of the examination of drawings and specification a first list of queries for the architect will be prepared. These should be written on A4 sheets on the left-hand half of the sheet and numbered serially, each batch also bearing a date. Ruled and headed sheets can be printed for the purpose at small expense. When a sufficient batch of queries has been collected, they can be typed and sent to the architect in duplicate, asking for the return of one copy with replies.

It may be more convenient for the surveyor to call on the architect to discuss the queries, in which case the replies can be noted, and on return to the office send another copy with replies filled in to the architect as confirmation of the decisions. This procedure can be repeated from time to time as necessary.

The queries should be given serial numbers, carried from one set to another, and the sets filed together as they are received, so that the series is complete. Each taker-off should, if possible, have a copy and mark off points as they are dealt with. A final check should be made in conference, when taking-off is finished, to see that there are no gaps or overlapping.

It will often happen that some proprietary material, of which the surveyor has never heard, is mentioned in the specification. If the name and address of the maker are not given, the surveyor should find them out and send for particulars of the material, which the makers are usually only too pleased to supply. At the same time an approximate price can be sought and this can be available should the question of cost of the material arise in discussion with the architect. In suitable cases it may be prudent to ask for a small sample, as the sight and handling of a piece of the material is often helpful in disposing of difficulties that may arise in describing the fixing.

A telephone enquiry to the Building Centre will almost always provide the name and address of the maker when only the trade name of a proprietary article is known. Lists of trade names with the names and addresses of manufacturers will be found in many reference books.

The surveyor should *never* accept an unknown name without investigation. A material specified has, before now, been found to be obsolete, and to require it in a bill of quantities not only looks silly and reveals ignorance but involves the raising of queries by tenderers. Even when the surveyor knows a material, if there has not been an occasion to refer to it for some time, up-to-date particulars should be sought. Specifications for use sometimes vary, new developments occur, and prices will fairly certainly have changed.

References to merchants' catalogue numbers or numbers of British Standards and Codes of Practice should be verified. A wrong figure may appear in the specification either through a typing error or through the writer not realizing that the reference is obsolete. Queries to an architect could take the form shown in Fig. 10.2.

Division of taking-off

For a small building it is most satisfactory if one person does the whole of the taking-off. For larger buildings the amount of subdivision will depend on the time allotted for the job and the availability of takers-off. Where two are made available a subdivision might be

A. Carcase of the building
B. Internal finish, windows, door and fittings.

Such sections as sanitary plumbing, drainage and roads are more or less independent of other sections, and could be allotted as one or other of the takers-off becomes available.

When three takers-off are available C could start on these sections, and on certain types of building involving a lot of joinery fittings, might take over fittings from B. There are few jobs except the very large ones in which it would be practicable to use more than four takers-off, and for the smaller jobs two or three would be normal.

The measurement of the carcase of a building would not normally be divided; one possible division would be into foundations and superstructure, and this

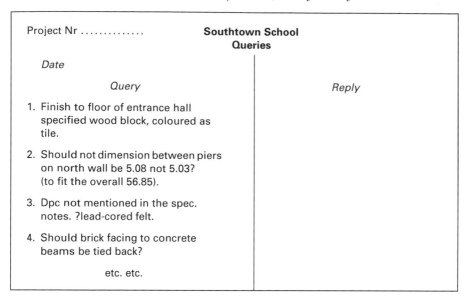

Fig. 10.2   Sample query list.

might be done if foundations are elaborate, or complicated by basements or pipe ducts. The superstructure might be subdivided for a steel or reinforced concrete frame building. The frame with floor, roof slabs and beam casings, would be the charge of one person, while the brickwork and roof coverings are dealt with by another. Such a section as roofs could, if necessary, be separated. If possible, however, one person should see the whole structure through.

In the same way windows and external doors can hardly be divided, as they are sometimes structurally combined and have similar finishings, and it would be preferable for internal doors to be done by the same person, as many items such as lintels, and plaster reveals, will occur in both sections. Internal finishings can be given to somebody else if a careful schedule has been prepared and there is close cooperation between the takers-off measuring finishings and openings.

The more the taking-off is subdivided the greater the risk of A thinking incorrectly that B has measured something, or C and D both measuring the same thing without each other realizing it. There are certain items in which practice differs with different offices. Some surveyors take skirtings with floor finish, others with wall finish. Some measure them net, others adjust for openings in the doors section. It is important, therefore, that a specific rule should be adopted on such items in the office, and care is necessary to see that new or temporary takers-off not used to the office custom are informed.

At set of 1:100 and 1:20 or other general drawings should be available for each taker-off, and it is worth the surveyor paying for the extra prints, if they are not supplied. Single copies would probably be sufficient of special drawings, such as details of joinery fittings or layout of plumbing services and drainage, as they would each in the main concern only one taker-off. If a second copy of the specification can be supplied it will be found to save a lot of time when two or more takers-off are engaged. On a large contract it may be worth while to have copies of the specification or substantial extracts photographed, particularly when, as may be the case, it is supplied in draft.

The amount of supervision of the taking-off necessarily varies, of course, with the experience of the individual. The junior taker-off just starting to do this work will need a good deal of watching and answering of queries. Cases have been known where one could almost do the work oneself in the time spent in supervision. However, the beginner must learn, and it must not be at the expense of serious mistakes. Quite a casual query may reveal unexpected ignorance, and with the inexperienced one must always be on the alert.

Computerized systems

These methods of measuring allow either the traditional process of 'writing' to the computer screen, or electronic measuring from computerized taking-off to the wider use of computer-aided design (CAD) systems. While such systems have existed in theory for a number of years (since the early 1980s), the industry is still some way from the pressing of buttons in order to generate automatic bills of quantities. In essence, because of their high initial and maintenance costs, such systems often only benefit the larger practices with the larger projects.

Electronic data interchange (EDI) also allows the process to be taken one step further by transferring the bills (or other documentation) directly into the contractor's computer for a form of 'automatic' estimating, using the contractor's own pricing information, which can then be analysed and adjusted to suit the particular project conditions (see also Chapter 4).

Buildings erected in stages

When buildings are to be erected in stages, separate prices may be required for each stage. If so, each stage will require an entirely separate bill or set of bills. It may be, however, that the division into stages is only for organization of the work: for example, when part of a building must be completed before another part can start. So far as the client is concerned only one price would be necessary, but it would be found of great convenience, both for certificate valuations and variation account, if each section were separately billed, provisional sums for specialist work being split up accordingly.

Similar circumstances arise for housing estates where houses are completed and handed over one at a time or in batches. Though it is not necessary to have a separate bill for each type, this does help. In any case, some idea of the value of each type must be arrived at, both for certificate valuations and for the release of retention, which is usually proportionally released for each house or batch of houses.

Taking-off for elemental bills

Where the billing is to be by elements instead of by trades, the taking-off must be subdivided accordingly. BCIS publishes a standard list of elements and this will usually suffice for most jobs. The peculiar character of the job may mean that all in the list cannot be used, and it may be that a new one must be introduced. In the interest of more accurate cost analysis the standard list should be adhered to so far as possible. If it is found practicable to make the sections of the taking-off correspond with the elements, it will probably be possible to save some working-up time.

Until one is used to taking-off by elements there is also the risk of something being missed. There is nothing to prevent the normal order and classification of taking-off being followed, so long as it is clear to the worker-up to which element each item is assigned. Each taker-off and worker-up on the job should have the list of elements, and it will be found useful to give code letters to each element, such as FDN for foundations, W EXTL for external walls or P for partitions. Dimensions can be marked up with these code letters in red before working-up begins.

Numbering dimension sheets

It is advisable to mark every page at the top with the section of the dimensions to which it belongs and the serial number of the page, such as Roofs 24 or Windows 38. On reference to any sheet one can then see quickly to what section it belongs. Sheets are also usually numbered serially right through, either on each page or on each column. If this latter can be done before abstracting is begun such numbering can be used for referencing in the abstract. If not, the sectional reference will have to be given. In either case it is important to ensure that sheets do not go astray, and it is advisable to keep a running index to the dimension sheets as a check, giving at the same time, whenever referred to, an overall view of what has been done of the taking-off. It need not at this stage be prepared in the full detail that is advisable in an index to be used for looking up dimensions, but it can form an outline for the fuller index.

Numbering on drawings

In a domestic building it may be practicable to describe each room by its name when referring to it in the dimensions, but in a building of any size it will be difficult to do so. Each room should then be given a serial number marked on all copies of the plans, and references made in the dimensions accordingly. In the same way, windows, doors and other openings can be given numbers in one or more series, so that a reference to the opening number in the dimensions will identify the window or door at once. The same references will be used in the schedules referred to above.

Alterations in taking-off

It often happens that alterations are made by the taker-off after the dimensions have been squared and perhaps at even a later stage when they have been carried through to the bill stage. It is most important that such alterations should be taken through all the stages of working-up. If the taker-off marks the alterations with a pencil cross and hands it personally to whoever is responsible for the working-up this should ensure that the corrections are made. If the working-up has been checked the pencil cross will not be rubbed out until the alteration has been checked, but it should then be erased, as otherwise every time it is seen it will raise a query as to whether something is left undone.

If the taker-off has made a number of alterations, a list of the page references on which the crosses appear can be handed over. Where the dimensions have not been worked up when altered, no further step will be necessary, except to cross out any relative squaring that has been done.

No alteration should be made to the taking-off, by a worker-up, supervisor or even the principal without reference to the taker-off, unless the person concerned is not available in time. There may be a reason behind the apparent error. Often, after making a correction, it turns out that the dimension was right the first time.

Standard descriptions

Most quantity surveying practices have developed their own style for their bills of quantities. To the casual onlooker all bills of quantities may look the same, but a more detailed examination will reveal personal differences and preferences. In an attempt to overcome misinterpretation by contractors and to ease the process of bill preparation, the use of standardized bill descriptions was developed. In addition, the effective use of computers for this purpose requires a standard library of descriptions. The bill items, instead of being written as a single description, are composed of a group of phrases. The descriptions are built up by using levels of phrases or words; by combining these, an ordered description is compiled.

Standard phraseology is not as easy to read or understand as traditional bill descriptions. A further disadvantage is that the work must be described in accordance with a standard library of descriptions, regardless of whether it could be better described by using other words, and there is the danger that inexperienced takers-off will opt for a standard description without giving due thought to the precise item description required.

Specialist bills

If for the purpose of prime cost sums it is required to get specialists' estimates based on the quantities measured, bills for this specialist work must be prepared first, to enable tenders to be obtained by the time the main bill goes to the printers, or at any rate by the time the proof is passed.

Such bills should state that the specialist will be a nominated subcontractor, and should state the form of main contract to be used. It is as well to mention the amount of cash discount to be allowed, as this may not agree with the specialist's normal practice. Similar bills can be prepared for the materials of nominated suppliers.

Bills should be sent in duplicate, so that the specialists have a copy of their instructions. They will sometimes quote on their own form, but it is advisable to ask for a copy of the bill to be signed and returned as confirmation of the term of enquiry.

Preliminary bill and preambles

If possible, before billing of the measured work starts, the preambles to each section covering materials and workmanship should be drafted from the specification, amplified if necessary by reference to previous bills. This should be done by one of the takers-off, who will by then have a comprehensive knowledge of the job. These preambles being written first, the biller can see how far the descriptions are already covered and make notes of any clauses which it is thought should be added.

The preliminary bill must also be drafted. If this is for a public authority, the surveyor may be supplied with their typical clauses. Otherwise the surveyor will probably follow a past bill. Care is necessary when using other bills to see that everything is applicable to the particular case. There may be clauses, parts of clauses, or even single words, which were inserted specially for the previous job and which, being in type like the rest, are not now evident as insertions. In the same way, owing to some particular circumstances, omissions may have been made previously that now need reinstatement. In the same way when standard preliminaries, either in-house, NBS or similar are used, care must be taken to ensure that they have been carefully completed and all gaps filled in and superfluous matter deleted.

As everything that concerns price must be in the bill of quantities, the unnecessary duplication of descriptions in specification and bill can be avoided by reference in the bill to clause numbers and headings of the specification, where there is a full specification prepared and issued with the bills. Where, however, the specification is not part of the contract, it is more satisfactory for the tenderer to have everything in the bill and save the extra time spent in cross-reference.

PC sums

Prime cost or PC sums are based on estimates received from potential nominated subcontractors. Such estimates can be obtained on request from the subcontractor concerned, who will have been appraised of what is required. While a submission of an estimate on its own is sufficient it is advisable to follow the procedures laid down by the JCT for nomination of a subcontractor. These procedures, as far as tendering is concerned, are as follows:

- NSC/T Part 1: the invitation to tender to be issued by the architect;
- NSC/T Part 2 the form of tender to be submitted by each subcontractor;
- NSC/T Part 3 the particular conditions to be agreed by contractor and subcontractor prior to entering the nominated subcontract.

and are subsequently followed by agreement, conditions, warranty and nomination forms.

Use of these procedures goes a long way to ensuring that the terms of the subcontract are clear and unambiguous, so greatly reducing grounds for dispute.

A check should be made of all PC items with the subcontractors' or merchants' estimates, with an eye to seeing that they include the proper cash discounts.

Where the proper cash discounts are not provided for, adjustment should be made so that the prime cost sum complies with the requirements of the contract. It may be noted:

- for converting net estimate to estimate subject to $2\frac{1}{2}\%$ add $\frac{1}{39}$;
- for converting net estimate subject to 5% add $\frac{1}{19}$;
- for converting net estimate subject to $2\frac{1}{2}\%$ to estimate subject to 5% add $\frac{1}{38}$.

Any incorrect discounts should be pointed out to the architect, so that the position may be regularized when acceptance is instructed. If the point is left and overlooked by the contractor, there may be difficulty with auditors if the adjustment is made in the variation account. Their view will be that the contractor should have examined the estimate and had it corrected before accepting it.

The surveyor must also ensure that provision is made for any materials to be supplied by the builder (for example cement and sand for tiling). There may be other special conditions accompanying a specialist's estimate, often set out in small print on the specialist's standard form. These must be examined to ensure that they are acceptable. For instance, there may be a requirement from a steelwork or structural floor contractor for a hard standing alongside the new building to allow hoisting direct from lorries into the final position. Requirements as to unloading vary, and these must be made clear.

It is advisable to mark the estimate with any adjustments made, so that the build-up of the sum given in the bill can be traced. As the estimate may not be retained it will be found useful to note on the dimensions against the provisional sum the name of the firm, date and reference of estimate, amount and discount. If there are any blanks in PC prices, provisional sums or quantities on the draft they should be marked in with a pencil cross so that they are not overlooked.

A schedule of all PC and provisional sums should be prepared showing how all the figures have been arrived at, and a copy should be sent to the architect in a form similar to that shown in Fig. 10.3.

Numbering items

Items in the bill can be serially numbered from beginning to end, either with a numbering machine if hand-drafted or automatically if computer-drafted. Alternatively, each page can have items referenced from A onwards through the alphabet, so that an item might be quoted as page 50 G or 94 B, for example. The advantage of this is that there is less risk of confusing item number and quantity, when quantity is on the left-hand side, but if the paper is ruled with an item column between double rules or in the form referred to below there should be little danger. However, the page number and lettering cannot be settled until the draft is in the typist's hands, so cross-referencing, often very useful to clarify the bill and help the estimator, is made more difficult. If a serial number is given in the draft that will be the same in the finished copy, the references can be finally filled in on the draft bill.

Southtown Primary School
Schedule of PC and provisional sums

Date

Item	Source of information, quotation		Amount included	Bill item no.
Nominated subcontractors		£		
Wood block flooring etc.	Messrs. Flooring Ltd's quotation ABC/JD dated... based on QS subcontract bill.	12 153.30		
	Add for mathematical mistakes in quotation	115.05		
		12 268.35	£12 500	34
Nominated suppliers				
Sanitary fittings etc.	Simple Sanitary Co. Ltd, estimate JA4163 dated ...	1 852.20	£1 890	48
Statutory undertakings				56
Water connection etc.	Southtown Water Board letter ABC/DEF dated ...		£200	
Provisional sums				
Contingencies Planting and sowing etc.	Spec. notes Item 18. Query No. 60		£2 000 £250	69 70

R S & T,
Chartered Quantity Surveyors.
Bank House,
Northtown.

Fig. 10.3 Specimen schedule of PC and provisional sums.

Note: Such a schedule might also be produced for named subcontractors under IFC 84.

Schedule of basic rates

If the contract alternative is used requiring recovery of fluctuations to be based on the rise or fall in prices of materials, an appendix will be required to the bill in which the contractor can set out the rates on which the tender is based. The appendix can contain a list of the principal materials (which the contractor can supplement if necessary) or it can be left to the contractor to prepare the list. Alternatively, a fixed priced list could be given of the principal materials, but it is preferable to let contractors put their own rates, as one contractor may be better placed than another through being a large buyer, or for some other reason.

If the priced bill is not being returned with the tender, it is advisable to reprint the schedule of basic rates of materials as an appendix to the form of tender as well, so that it can be considered when tenders are compared.

Schedule of allocation

When the price adjustment formula for calculating variation of price claims is to be used it is necessary when the bill of quantities is complete, and before it is sent out to tender, to prepare an allocation of all the items in the bill. This is done by preparing a schedule of all 49 work categories and then going through the bill allocating the items to the work categories or, as for preliminaries and certain provisional sums, into a 'balance of adjustable items' section.

When this allocation is complete it is bound in the back of the bill, and in submitting a tender a prospective contractor is deemed to have agreed to the allocation chosen by the quantity surveyor. It is unlikely that a tendering contractor will challenge the allocation, but if they do they must do so when submitting the tender. When the tender has been accepted it is a straightforward exercise to complete the schedule by filling in the money; the whole thing is self-checking in that the end result must be the contract sum.

Dispatch to printers

The draft bill should receive a final careful reading through, an overall checking of the quantities and editing before it is sent to the printers or word-processed in-house, and an examination should be made to see that all blanks have been filled in, such as cross-references to item numbers or summary pages. If possible it should also be looked through by the principal or a senior assistant other than the takers-off, who at this stage quite possibly cannot 'see the wood for the trees'. An outsider will often be able to put a finger on something wrong.

Careful examination should also be made of the drawings and specification. All notes on the drawings should be looked over in case any have been missed. 'Of course, you've taken this' may produce the answer, 'No, I thought *you* had!' The specification can be run through in pencil, clause by clause, in confirmation that each has been covered, either by the taker-off measuring or by whoever drafted the preliminary bill or preambles.

At some stage between completion of the draft bill and passing of the proof the whole of the dimensions and abstract should be looked through to see that

- all squaring is checked;
- all casts and reducing on abstract or cut slips are checked;
- all items in the draft bill or cut slips are checked.

If the surveyor is arranging for printing, the printers' instructions as to number of copies and binding should be clearly made in writing. In addition to the number of copies required for tenderers, allowance must be made for additional copies. These vary with circumstances but might be as follows:

Contract	2
Client	1
Architect	1
Quantity surveyor	2
Successful contractor	2
Clerk of works	1
Spare	2 or more

'Spare' may be required for a last-minute addition to the list of contractors or for, perhaps, the freeholder or other person to whose approval the work is to be carried out.

Invitations to tender

The project team will usually prepare, often in consultation with the client, a list of firms to be invited to tender, in some instances by way of some form of prequalification such as interview. The firms chosen to be on the list will be sent a letter of invitation; on receipt of replies a list of those who have accepted can be prepared. Such a letter should include at least:

- name of client and architect
- title and location of the job
- approximate date when bills of quantities will be issued
- time to be allowed for tender
- where the drawings may be seen or some description of the works
- the form of contract to be used.

A typical letter for such a case is given in Fig. 10.4, based on the example published in the NJCC Code of Procedure for Selective Tendering (obtainable from RICS). Occasionally some of the requirements of the code of procedure are not necessary; the requirement, for instance, that tenderers should have two copies of the bill is important if they must send a priced copy with their tender or even forward it on advice that their tender is under consideration. If, however, a blank copy is sent after receipt of tenders (two or three days being allowed for filling in the prices), the extra expense of two copies to all tenderers does not seem justified.

The offer of additional copies of the bill, or of sections of it, which the above-mentioned code stipulates is only likely to be necessary in large contracts.

Form of tender and envelopes

Except for clients who have their own standard tender forms, the surveyor will probably be required to draft one to go to the printers with the draft bill. Envelopes of suitable size to hold the tender form will also be required, addressed to the architect or according to the arrangements made for delivery of tenders, and marked with the name of the job so that they can be put aside on receipt and all opened together. The number of tender forms will be twice the number of tenderers (if each is being sent two copies) with, say, two or three extra copies, one of which should be sent to the architect for reference.

If the contractor is to be given the option to tender for work covered by prime cost sums, in accordance with clause 35.2 of JCT 80, it is advisable to make provision in the form of tender for tenderers to state what work (if any) they would like to tender for. With the current emphasis on early planning, the architect may want to settle some of the specialists provisionally when preparing the drawings, and so obtain the benefit of consultation with them. Since this contract clause leaves the matter entirely to the discretion of the architect, it has little practical value in such cases.

Some public authorities require the priced bill to be returned with the tender, but this is not usual, though sometimes done, in private practice. If it is required, a separate addressed envelope for return should be provided, of suitable size

Dear Sirs

Southtown Church of England Primary School

We are authorized to prepare a preliminary list of tenderers for the construction of the works described below.

Your attention is drawn to the fact that apart from the alternative clauses to the Standard Form of Building Contract as detailed below under item j, further amendments to the Standard Form of Contract are annexed hereto and will be incorporated in the tender documents.

Will you please indicate whether you wish to be invited to submit a tender for these works on this basis. Your acceptance will imply your agreement to submit a wholly bona fide tender in accordance with the principles laid down in the 'Code of Procedure for Single Stage Selective Tendering', and not to divulge your tender price to any person or body before the time for submission of tenders. Once the contract has been let we undertake to supply all tenderers with a list of the tender prices.

Please state whether you would require any additional unbound copies of the bills of quantities in addition to the two copies you would receive; a charge may be made for extra copies.

You are requested to apply by *date*. Your inability to accept will in no way prejudice your opportunities for tendering for further work under our direction; neither will your inclusion in the preliminary list at this stage guarantee that you will subsequently receive a formal invitation to tender for these works.

Yours faithfully

a   Job: Southtown Church of England Primary School
b   Employer: Blankchester Diocesan Board of Finance
c   Architect: LMN Chartered Architects
d   Quantity Surveyor: RS&T Chartered Quantity Surveyors
e   Consultants: None
f   Location of site: Site plan enclosed
g   General description of work: New Primary School
h   Approximate cost range: £700–800 000
i   Nominated subcontractors for major items: Engineering Services
j   Form of Contract: JCT 80 incorporating amendments 1–12
      Clause 15.2: VAT clause will apply
      Clause 19.1.2: will apply
      Clause 21.2.1: insurance is not required
      Clause 22A: will apply
      Clause 23.1: will not apply
      Clause 38: will apply
      Clauses 41.2.1/41.2.2: will not apply
k   Percentage to be included under Clause 38.7: 10%
l   Examination and correction of bills: Alternative 1 will apply
m   The contract is to be: under hand
n   Anticipated date for possession: *date*
o   Period of completion of the works: 65 weeks
p   Approximate date for dispatch of all tender documents: *date*
q   Tender period: 5 weeks
r   Tender to remain open for: 4 weeks
s   Liquidated damages value: £1000 per week
t   Details of Bond requirement: None

Fig. 10.4   Preliminary enquiry for invitation to tender.

Source: Based on the form suggested in the *Code of Procedure for Single Stage Selective Tendering*, Appendix A1, NJCC.

and strength to hold the bill. The priced bill will be delivered sealed, and only opened if to be considered for acceptance: its envelope must, therefore, be marked with the name of the job in the same way as that enclosing the tender form, and the tenderers should be instructed that each must put their name clearly on the outside. See Fig. 10.5.

Reading proofs

Whatever method of duplicating is used a proof should be supplied. The reading through of a proof is a laborious task. One person comparing draft and proof will very easily miss differences, and probably the best way is for two to check, one reading from the draft and the other following in the proof. Periodically the duties and documents should be changed to avoid the soporific effect of listening

---

TENDER FOR: Southtown Church of England Primary School

TO: LMN Chartered Architects

Sirs

We having read the conditions of contract and the bills of quantities delivered to us and having examined the drawings referred to therein do hereby offer to execute and complete in accordance with the conditions of contract the whole of the works described for the sum of £726 547.00 and within 65 weeks from the date of possession.

We agree that should obvious errors in pricing or errors in arithmetic be discovered before acceptance of this offer in the priced bills of quantities submitted by us, these errors will be dealt with in accordance with Alternative 1 of the Code for Single Stage Selective Tendering.

This tender remains open for 28 days from the date fixed for the lodgement of tenders.

Dated this .... day of .... 19......

Name ....

Address ....
        ....
        ....

Signature ....

Fig. 10.5   Form of tender.

Source: Based on the form suggested in the *Code of Procedure for Single Stage Selective Tendering*, Appendix C: NJCC.

Note: If used in Scotland then there is a need to add a paragraph confirming the tenderer's willingness to enter into a formal contract, and the signature requires two witnesses.

to the reading for too long. A good way to simplify the reading through is to go through the quantities columns first to ensure

- correctness of figures, both of item numbers and quantities;
- that they are in the right column;
- that the 'm' or 'm^2' or 'm^3' is correct.

Descriptions can then be read through separately. This method is particularly valuable if only one person is reading through the proof, as in the usual way one is apt to relax concentration on the three above-mentioned points, which are all-important. In all cases reading should be from the draft, as being slower than reading from type: otherwise the reading would be too fast to follow properly.

Issue of drawings

It is necessary to issue to tenderers a copy of each of the location drawings: the site plan, the general 1:100 sheets, and certain of the component details in accordance with the requirements of SMM. Sufficient copies of these should be obtained from the architect in time for issue with the bills.

If often happens that in preparation of the bill of quantities errors are found in the architect's drawings, which should be corrected. The architect should be advised of these in time to correct the originals before prints are taken for issue to tenderers.

Dispatch of finished bills

A covering letter (see Fig. 10.6) must be drafted to go to tenderers with the bills. It should state:

- what documents are enclosed;
- date, time and place for delivery of tenders, and that tenders are to be delivered in the envelope supplied;
- what drawings are enclosed and where and when further drawings can be seen;
- what arrangements the tenderer must make for visiting the site: with whom appointment should be made, or where the key can be obtained. If the site is open for inspection this should be stated;
- a request for acknowledgement.

It is also advisable to state that the client (employer) is not bound to accept the lowest or any tender or to pay any expenses incurred by the tenderer in preparing the tender.

Care must be taken in arranging the documents for each contract, so that all have them complete and correct. They are probably best laid out in piles with their envelopes and checked as they are put into them.

Date

Dear Sirs

Southtown Church of England Primary School

Following your acceptance of the invitation to tender for the above, we now have pleasure in enclosing the following:

a two copies of the bills of quantities;

b two copies of the general arrangement drawings indicating the general character and shape and disposition of the works, and two copies of all detailed drawings referred to in the bills of quantities;

c two copies of the form of tender;

d addressed envelopes for the return of the tender and priced bills of quantities together with instructions relating thereto.

Will you please also note:

1 Drawings and details may be inspected at the offices of the architect.

2 The site may be inspected by arrangement with the architect.

3 Tendering procedure will be in accordance with the principles of The Code of Procedure for Single Stage Selective Tendering.

4 Examination and adjustment of priced bills (Section 6 of the Code) Alternative 1 will apply.

The completed form of tender is to be sealed in the endorsed envelope provided and delivered or sent by post to reach the architect's office not later than 12.00 noon on *date*.

Please acknowledge receipt of this letter and enclosures and confirm that you are able to submit a tender in accordance with these instructions.

Yours faithfully

Architect

Fig. 10.6   Formal invitation to tender.

Source: Based on the form suggested in the *Code of Procedure for Single Stage Selective Tendering,* Appendix B: NJCC.

Note: In Scotland it is mandatory for the priced bills to be returned with the tender; elsewhere it is a desirable alternative to the option of delivery within four days.

Correction of errors

Once the bills are dispatched, a copy should be looked through, as mistakes, made perhaps in the rush to get the bill off in time, may catch the eye. Queries may arise, too, from contractors tendering. Unless errors are of a very minor nature, they should be circulated to all contractors in time for them to correct their copies before tenders are made up. An acknowledgement should be requested to ensure that all tenderers have had the opportunity of making the corrections. Even so, when the priced bill of the successful contractor is being examined, it should be verified that the corrections have been made. A typical letter will be found in Fig. 10.7.

Date

Dear Sirs,

Southtown School: New Extensions

Will you please make the following corrections in the bill of quantities:
Item 246. For 16 m^3 read '66 m^3'.
    356. For 'm^2' read 'm'.
Please acknowledge receipt of this letter.

Yours faithfully,
R S & T

Fig. 10.7 Circular letter to contractors (corrections to the bill of quantities).

Copyright in the bills of quantities

A word should be said on the subject of copyright. Copyright is established by the Copyright Act 1988 in every 'original literary, dramatic, or musical work'. It is not clearly established whether there is copyright in a bill of quantities. The RICS took the opinion of counsel, and were advised that in his opinion copyright existed, on the ground that the bill was an original literary work within the meaning of the Act of 1911 which used the same wording. But this is only an opinion and, as has already been pointed out, opinions differ, and by so differing produce defended actions at law. There are inevitably many clauses in a bill of quantities that are more or less standard and used in very similar form by many surveyors, but the quantities are undoubtedly original, and there will be in all bills a number of items that are original and peculiar to that particular bill.

As there might be those who would take advantage of an opportunity to reuse a surveyor's bill, the best protection would appear to be specific reservation by surveyors in their agreements with their clients of the rights of reprinting and reusing bills. If, of course, the surveyor is advised, when instructed, that it is proposed to use the bill again as and when required, and the condition is accepted, there can be no complaint. The difficulty is avoided if the RICS Form of Agreement, Terms and Conditions for the Appointment of a Quantity Surveyor is adopted unamended, as copyright in the bills of quantities is retained by the quantity surveyor in clause 8 of that agreement.

The dimensions, abstract and other memoranda from which the bill of quantities is prepared are in a different category. These are the surveyor's own means to attain an end: the bill of quantities which contractually has to be provided. It has been held that the surveyor is entitled to retain these documents (unless, of course, as is sometimes the case, the contract with the client provides otherwise).

Specifications and schedules of work

In the context of tender and contract documentation the specification has always played a key role. With the introduction of CPI, as described above, the specification has become more important than ever. BPIC in their publications make it clear that the specification is the key document from which all other information, either for drawings or bills of quantities, will flow.

The writing and use of specifications is a subject in its own right and as such warrants separate study. Suffice it therefore in this book to restrict comment to what a specification is for and the changes that have come about in recent years in the way that specifications are drafted.

Drafting specifications

For many years it was common practice for specifications to be handwritten, albeit often using previous documentation suitably amended. Over the intervening years the practice of writing specifications fell into decline. On many occasions specifications became a matter of a few sheets of hastily drafted notes; more often it was a case of 'It's all on the drawings'.

Today, owing to the advances in computer technology, slowly at first but with gathering momentum, standard specifications have become the order of the day. Now architects and surveyors can enjoy the benefits of having the facility to call up mark-up copies of a standard specification to be adapted for each specific project.

National Building Specification

The National Building Specification (NBS) is not a standard specification; rather, it is a large library of specification clauses, all of which are optional. Many are direct alternatives, and often require the insertion of additional information. NBS thus facilitates the production of specification text specific to each project, including all relevant matters and excluding text that does not apply.

NBS is available only as a subscription service, and in this way it is kept up to date by issue of new material several times a year for insertion into loose-leaf ringbinders. NBS is prepared in CAWS matching SMM, and complies fully with the recommendations of the CPI Code of Procedure for Project Specifications. There are three versions of NBS, the Standard Version, an abridged Intermediate Version and a Minor Works Version.

Schedules of work

A schedule of work is a list of items of work required to be done and should not be confused with a specification. It is mainly used in works of alteration to spell out the items that are only covered in the specification in general terms. Recently, schedules of work have started to appear as an adjunct to the specification, but they have to be used with care.

A specification, like a bill of quantities, incorporates contract particulars, client's requirements and contractor's liabilities, as well as a full specification of the materials and workmanship. A specification, however, should never contain quantities; these are matters for the quantity surveyor when there are bills and for the contractor when there are no bills.

To quote quantities in a specification is inviting trouble: the contractors will say, 'We've priced the quantities we were given' whereas they should have priced everything that they considered necessary from their own measurements to arrive at a lump-sum price. Where the specification option is used in some standard forms of contract, the inclusion of quantities can lead to those items gaining priority over the drawings.

In the same way problems can arise when the description of the work is set out in schedule form following the materials and workmanship clauses. Old-fashioned specifications used to end up with words such as 'Carry out all the work shown on the drawings'. Today there is a tendency on the part of clients to require the lump sum to be broken down into component parts, and schedules of work are appearing indicating specific packages, such as alterations, substructure, brickwork and roofing, and a £ sign is appearing against each of these packages.

While this can be of some assistance in checking interim valuation applications, in giving the client a breakdown of the price and in some ways costing variations, the same problem exists: 'We only priced what was written down', whereas the intention was that they should have priced everything necessary. Because of these problems the same care must be taken in drafting these schedules of work as is taken in drafting the specification itself. It must be very clear to the estimator exactly what is wanted, and nothing must be missed.

Other documents

As mentioned at the beginning of this chapter, bills of quantities have traditionally been the mainstay of a quantity surveyor's work. This is now no longer necessarily the case. While bills of quantities remain an important part of their work, with the advent of other forms of procurement as described in Chapter 9 the quantity surveyor has become involved in the preparation of other documents as well.

Management contracting requires the preparation of work packages, each to be let to different subcontractors. These may be produced by the client's quantity surveyor or by the contractor's own in-house quantity surveyors. Whoever does it, it must be done with care to ensure that all relevant information is provided.

With the onset of design-and-build contracting there is a need for someone to draft the enquiry document in either one stage or two: a task which often falls to the client's quantity surveyor. Perhaps the most important element of these documents is the employer's requirements, and great care must be taken to ensure that all such requirements are sought and carefully considered and comprehensively drafted. These employer's requirements will form the basis of the contractor's tender, and everything that is required which is not contained within the employer's requirements will be considered and dealt with by the contractor's proposals and so are outside the control of the client and their representatives. Variations can only be made by way of an amendment to employer's requirements, and it is therefore important to ensure that they are fully comprehensive.

Receipt of tenders

Delivery and opening

In public authorities tenders will probably be addressed to the secretary or principal chief officer, but with private clients they are usually forwarded to the office of the architect or quantity surveyor. When the time for delivery comes the envelopes received will be counted, as a check that all are in, before being opened. After opening, the official concerned or the architect or quantity surveyor, as the case may be, will prepare a list of the tendered amounts, have them arranged in order and typed for reporting.

The practice sometimes adopted of not giving contractors the list of tenders, or giving the figures only without the names, is to be discouraged. Publication of the result is the least that can be done in return for the time and trouble taken by tenderers without charge. It is advisable that tenders delivered after the time fixed should be rejected, since after the published time of opening contractors are apt to telephone each other as to figures, and an unscrupulous contractor, if there were such a one, might take advantage of this. If the postmark showed the tender to have been dispatched before the time for delivery, it might well be accepted. An alternative, if the lower tender is late, is to notify all tenderers that the time for receipt of tenders has been extended by, say, seven days, and ask all to confirm their tenders.

Reporting of tenders

In considering tenders other factors than the tendered price may be of importance. The time required to carry out the work, if stated on the form of tender, may be compared, as time may be very important financially to the client. Although there may be reasonable excuses for failing to keep to the time arranged, and even justification for avoiding the liquidated damages provided for by the contract, the time stated by a reputable contractor may be taken as a reasonable estimate, having regard to the prevailing circumstances.

If the contract is subject to adjustment of the price of materials, the schedule of basic rates of materials must also be considered and the question asked: Has the tenderer assumed reasonable basic prices for materials? If they have been based too low there will be an excessive 'increased cost' on a rising market or too little in 'reduced cost' on a falling market. Where tenders are close the schedules of basic rates should be compared, as the lower tenderer may have less favourable rates. Only a preliminary examination will be made at this stage to ascertain which tender or tenders should be considered for acceptance; a fuller report, after supporting estimates are produced, will be made later by the quantity surveyor.

The architect, having considered these matters in consultation with the quantity surveyor, or the public authority in consultation with both the technical advisers, will report the tenders to the client or committee concerned and set out clearly for their consideration the arguments in favour of acceptance of one tender or another, if there is any doubt.

When tenders are invited from a limited number of contractors the lowest, or potentially lowest, should be accepted. All go to a good deal of trouble and expense in preparing a tender, and the object of such tendering is to decide which amongst a number, all acceptable to the building owner, will do the work at the lowest price. Whether expressly disclaimed in the invitation or not, there is no legal obligation to accept the lowest or any tender, but there is in a limited invitation a moral one to accept the lowest, if any.

When tenders are advertised and any tenderer who can find the required deposit and surety may submit a tender, the circumstances are different; there might be justification for rejecting the lowest tender on grounds other than cost. However, when the expenditure of public money is involved, there may be repercussions and the grounds for such rejection will need to be clearly demonstrated.

Examination of priced bill

Before acceptance of a tender, the tenderer whose offer is under consideration will be asked to send a copy of the priced bill to the quantity surveyor for examination, if it has not been delivered with the tender, a blank copy being sent for the purpose. Sometimes, to save time, the original bill is called for but a copy is sufficient. The original may be marked with estimator's notes which it would be injudicious to disclose. There is no justification for the certificate one has sometimes seen asked for that the copy of the bill has been compared and checked with the original. The tender is a lump-sum tender and the sole purpose of obtaining the pricing is to provide a fair schedule for the adjustment of variations.

In the first place a check will be made of all mathematics, so that if clerical errors have occurred it is quite clear what rate shall be used for adjustment. If, for instance, an item has been priced at £0.50 per m and extended at £0.05, it will not be fair that either additional quantity or omission of the item should be priced out at the incorrect rate in adjusting accounts. All clerical errors should be corrected in the contract copy of the bill but, of course, the amount of the tender would not normally be altered. The difference will be shown as a rebate or plusage in the summary (see Fig. 10.8) which will be applicable, in interim valuations and adjustment of variations, to all rates except provisional sums and PC prices.

Apart from the arithmetical check a technical check should be made of the pricing by generally looking through the rates. A slip may be noticed of an item left unpriced, or one that is billed in square metres being priced at what is obviously a linear metre rate, or vice versa. Or some obvious misunderstanding of a description may be noted. Here again, corrections should be made so that a reasonable schedule of rates for pricing variations results.

A secondary reason for examination of the priced bill is to ensure that the tenderer has not made such a serious mistake that they would prefer to withdraw the tender, as they may at any time before acceptance. If such a serious error is found out, it is advisable, subject to the architect's approval, to draw the tenderer's attention to it, otherwise, finding it out sooner or later, there is a risk that constant attempts will then be made to recover the loss, possibly to the detriment of the client's interest. If correction of the error does not bring the

Summary	**£**
Preliminary items	18 662.00
PC and provisional sums	55 408.00
Substructure	17 479.06
Concrete work	8 131.16 ~~8 031.16~~
Brickwork and blockwork	19 083.69
Roofing	4 789.88
Woodwork	18 103.97
Metalwork	4 392.36 ~~4 393.36~~
Plumbing services	1 884.38 ~~1 884.28~~
Plasterwork and finishings	8 616.01
Glazing	1 036.77
Painting	4 302.74
Drainage	3 728.98
External works	16 445.98
	£182 064.48 ~~£181 930.68~~
Water and insurance 3.5%	6 367.57
	£188 432.25 ~~£188 298.25~~

Tender submitted £188.250.00

In the above summary alterations have been made correcting clerical errors etc. assumed to
have been found in the priced bill. The increased total means that all rates (except PC and
provisional sums, which the contractor has no power to reduce) will be subject to a percentage
rebate. To find this, extract of PC and provisional sums will first be made as follows.

	£
Provisional sums	10 000
Mechanical services	14 850
Electrical services	8 223
Wood flooring	2 200
External staircase	4 050
Metal windows	6 500
Water mains	200
Ironmongery	550
Sanitary fittings	890
Daywork	2 610
	£50 073

The rebate to be expressed as a percentage is:

	£	£
Errors	182 064	
	181 930	
		134
Rebate in tender	188 298	
	188 250	
		48
		£182

(This total equals the total difference between £188 432 and £188 250.)

The percentage is worked out as follows:

	£
Corrected total without insurance, etc.	182 064
Less PC and provisional sums	50 073
	£131 991

$$\text{Percentage} = \frac{182}{131\,991} \times \frac{100}{1} = 0.14\%$$

In the variation account, therefore, all rates will be subject to addition of 3.50% for water and insurances, and all except PC and provisional sums and accounts set against these will also be subject to a rebate of 0.14%. The water and insurance percentage can be converted into a percentage on the contractor's own work instead of on the whole total as appears in this example, and in that case the two percentages can be combined into a single percentage. As, however, in this case the contractor has expressed water and insurances as a percentage of the whole, they are so treated.

Fig. 10.8 Summary corrected.

tender under consideration above the next highest, the architect may feel that the client should be advised to allow amendment of the tender. Otherwise the next tender falls to be considered. If a tender has been accepted and a contract formed, tenderers strictly cannot withdraw, but it may not be policy to hold them to their tender for the above-mentioned reason.

If priced bills are delivered with the tender by all contractors, only the bills of tenderers under consideration for acceptance should be opened, usually one, but sometimes, if tenders are close, two or three. All others should be returned unopened.

Correction of errors

Under section 6 of the Code of Procedure for Selective Tendering, 'Examination and Correction of the Priced Bills', two alternatives appear for dealing with genuine errors. Under alternative 1 the contractor either confirms or withdraws, the situation that previously existed and which is dealt with in this book. The second alternative allows the contractor to confirm or correct. The difficulty is to decide what is a genuine error, particularly as, under 5.2, four days can elapse before the priced bill of quantities must be produced (unless the option of return with tenders is adopted). Whichever alternative is to be adopted, it is important that a decision is made before tenders are invited and that such decision is communicated to the contractors tendering.

Examination of schedule of basic rates

Where a contract is to be subject to price adjustment, and the formula method is not being used, when the copy of the priced bill of quantities is asked for from the contractor whose tender is being considered, they should at the same time be asked to submit estimates in support of the basic prices for materials on which they require adjustment (if any). Any rates not the subject of specific quotation must be carefully checked. Certain materials, such as Portland cement, have standard prices; others, such as steel tubing or stoneware drain goods, are quoted at a percentage on or off a standard list. The current rates can be obtained from the trade association concerned. The bulk of the basic rates will be probably be arrived at from ad hoc estimates from merchants.

Note that prices will often vary with the size of the 'lot' quoted. A basic price for cement quoted in 10 tonne lots should be so marked; if it is bought in smaller lots, adjustment will be against the corresponding basic price of that size lot.

Quotations should be examined with a knowledge of prices in mind. It is not unknown for merchants to make mistakes against themselves in the price quoted. It will probably be some months before the quotation is accepted, and then only after new tenders have been obtained by the contractor in an attempt to get a lower one. The mistake being discovered, the contractor may claim the correction as increased cost. Under the normal forms of contract fluctuation in market price must be proved, so recovery of the amount of the error cannot be made.

The introduction of the price-adjustment formula negates this process and instead the quantity surveyor will be required to complete the schedule of allocation and the reduction of the work categories to the agreed number if required.

Reduction bills

If, as often happens, the lowest tender is for a higher sum than the client is prepared to spend, it may be possible to bring it within the financial limit by making some adjustments on the work required. This would be done by preparing an adjustment bill, incorporating the revisions and so revising the tender, in a similar way to the variation account referred to in Chapter 11. As this bill is prepared before the contract is signed, the contract sum would be the revised tender figure. The bill, in fact, modifies the original quantities, and the quantities so modified become part of the contract. The adjustments will probably be mostly omissions, but if balancing additions are required, for which there are no prices in the bill, the prices for these would have to be agreed with the contractor.

Sometimes, if a variation is complicated but its value can be estimated fairly accurately, the adjustment could be made as a lump sum at this stage, either as a sum agreed by the contractor or as a provisional sum subject to adjustment in the variation account.

In the same way any afterthoughts or additional requirements could be made the subject of an adjustment or addenda bill, so that they are incorporated in the contract before signature.

Preparation of contract bill of quantities

A fair copy of the priced bill of quantities will be required for signature with the contract. If a blank copy has been sent to the contractor to be completed, this can be used as the surveyor's office copy after making any alterations necessitated by the check, but a corrected copy for the contract is probably best made in the surveyor's office either by hand or by photocopying the original. If made by hand it should, of course, be in ink and contain the schedule of basic rates of materials or the allocation of the items, so making them part of the contract. It will then be sent to the architect, or quantity surveyor, or clerk in the case of a public authority, who will be getting together the contract documents.

The prices in the contractor's priced bill of quantities must be treated as confidential. They must be used solely for the purpose of the contract (see JCT 80, clause 5.7). Though they naturally contribute to the surveyor's knowledge of current rates, they can only be referred to for the surveyor's own information, but not for example when discussing a price with a contractor, say 'X's price on such-and-such job was so much.' Where the contractors have been required to deliver their priced bills with the tender, those not considered for acceptance being returned unopened, their prices are not disclosed to anybody. (It is not unknown for contractors to submit tenders out of courtesy, because they do not like to refuse, and in such a case the blankness of the 'priced' bill will also be kept concealed by this procedure.)

Preparing the contract

The duty of preparing the contract by completing the various blanks in the articles of agreement falls on the architect although the quantity surveyor is often asked to do it. It may be necessary to add special clauses to the conditions of contract and to amend other clauses; if so, they must be written in, and the insertion or alteration must be initialled by both parties at the time of signature. Extreme care must be taken if clauses within standard conditions of contract are to be amended, as specific alterations can affect other clauses. It is always advisable to seek legal advice if it is intended to make substantial amendments. Any portions to be deleted must be ruled through and similarly initialled. All other documents contained in the contract (each drawing and the bill of quantities) should be marked for identification and signed by the parties; for example,

> This is one of the drawings
> or   This is the bill of quantities
> referred to in the contract
> signed by us this   day of   19

For the bill of quantities this identification should be on the front cover or on the last page, and the number of pages can be stated. If the standard form is used (with quantities) the specification as such is not part of the contract, and will not be signed by the parties. If there are no quantities the full specification is a contract document and must be signed accordingly. All the signed documents must be construed together as the contract for the project.

Contracts are either signed under hand, when the limitation period is six years or, when 12 years is required, are completed as a deed (formerly under seal). In the latter case it is important to ensure that this is duly recognized, as failure to do so could have serious implications.

Case law exists that illustrates the importance of ensuring that all the contract documents are in agreement with each other. In *Gleeson* v. *London Borough of*

Hillingdon (1970) EGD 495 there was a discrepancy between completion dates set out in the contract bills and the completion date in the appendix to the JCT form of contract. Delays had occurred, and the question was which date was to be taken in calculating liquidated damages. The court held that under the relevant clause of the standard form in use at that time (JCT 63 clause 12(1)) the date in the appendix prevailed, but the litigation would not have occurred if all the contract documents had been checked for inconsistencies.

Bibliography

Allott, A. *et al. Common Arrangement for Specifications and Quantities.* CCPI, 1984.
Burrows, M. 'Tendering documentation 1750–1850'. *Chartered Quantity Surveyor*, December 1982.
Daly, I.M. 'What's the use of bills of quantities?' *Building Economist*, September 1981.
Dearle and Henderson *SMM7 Reviewed*. Blackwell Scientific Publications, 1988.
Harmer, S. 'Identifying significant BQ items'. *Chartered Quantity Surveyor*, October 1983.
Hibberd, P. 'The case for analytical bills'. *Chartered Quantity Surveyor*, October 1982.
Horne, R. 'CESMM and SMM6: the differences'. *Chartered Quantity Surveyor*, November 1983.
JCT *Revised Procedure for Nomination of a Sub-Contractor*. Joint Contracts Tribunal. 1991.
Nisbet, J. 'Identifying the knowledge base'. *Chartered Quantity Surveyor*, June 1991.
NJCC *Code of Procedure for Selective Tendering*. National Joint Consultative Committee, 1977.
RICS *Appointment of a Quantity Surveyor*. The Royal Institution of Chartered Surveyors, 1983.
SCQSLG *The Presentation and Format of Standard Preliminaries*. Society of Chartered Quantity Surveyors in Local Government, 1981.
Skinner, D.W.H. 'The contractor's use of bills of quantities'. CIOB Occasional Paper, Chartered Institute of Building, 1981.
The Core Skills and Knowledge Base of the Quantity Surveyor. RICS Research Paper No. 19, The Royal Institution of Chartered Surveyors, 1992.
Townsend, G. 'Industrial engineering – an acceptable method at last'. *Chartered Quantity Surveyor*, July 1985.
Trickey, G. 'Why have a bill of quantities?' *Architects Journal*, July 1982.
Turner, W. 'Dispense with BQs at your peril'. *Chartered Quantity Surveyor*, 1981.
Willis, C.J. and Newman, D. *Elements of Quantity Surveying*. Blackwell Scientific Publications, 1992.
Willis, C.J. and Willis, J.A. *Specification Writing for Architects and Surveyors*. Blackwell Scientific Publications, 1991.
Willis, C.J. and Moseley, L. *More Advanced Quantity Surveying*. Granada, 1983.
Wilshere, D. 'Bills widespread in engineering services'. *Chartered Quantity Surveyor*, September 1982.

Chapter 11

Post-contract procedures

Introduction

The scope of the quantity surveyor's involvement during the post-contract stage of a project will generally fall under the following headings:

- valuations
- cost control/reporting
- final accounts
- claims.

Depending upon the size and nature of the project, the post-contract administration may be undertaken by site-based staff or office-based staff involved on a full-time or intermittent basis. Nevertheless, the duties to be performed will remain similar.

The degree of involvement may vary according to the type of main contract. For example, if the contract is awarded on an approximate quantities basis requiring remeasurement on site, there will be a need for additional surveyors to be involved to carry out the site measurement.

The successful execution and completion of the post-contract procedures, and the final account in particular, are very much dependent upon cooperation between the client's appointed quantity surveyor and the contractor's surveyor. There are areas of involvement common to both, such as allocation of work categories for final account purposes, registration and allocation of instructions and drawings, agreement of site measurements and the early resolution of dayworks. By early planning and agreement these issues can be addressed to the benefit of all concerned.

Valuations

In the course of a major building contract an architect or contract administrator is invariably called upon to make decisions and to issue a series of certificates which must be issued in accordance with the provisions of the particular contract. In so doing when acting in an arbitral situation, an architect is under a duty imposed by law to act fairly and impartially between the parties.

Where a valuation certificate is required by the contract, and certainly under JCT 80, its issue is a condition precedent to payment. If the certificate is improperly withheld, particularly if the delay arises from any action by the employer, entitlement to payment may be enforced in the absence of the requisite certificate.

Accuracy

The valuation for certificates should be made as accurately as is reasonably possible. The contractor is entitled under the contract to the value of work done, less a specified retention sum. If the valuation is kept low the retention sum is in effect increased. To a contractor having a number of contracts in hand these excessive margins will mount up and demand additional capital. However, the client must be protected against the possible insolvency of the contractor, when an overpayment resulting from an excessive valuation cannot be recovered, and the client might have to meet extra expense in employing another contractor to complete the work.

Dates

If, as is often the case, the contractor is entitled to certificates monthly, it will be found convenient to arrange the dates at the beginning of the contract: say, the last Thursday in every month. Sometimes the date must be fixed to suit the client's convenience, particularly when payment is passed at a board or committee meeting held at fixed intervals, as every effort should be made to reduce the interval between valuation and payment. It helps if the contractor submits a valuation statement, but the responsibility is that of the surveyor. A meeting should be arranged with the contractor's surveyor, as it is an advantage for the valuation to be mutually agreed.

Nominated subcontractors should be notified by the main contractor of the dates of valuations and be required to submit statements by those dates. If preferred, the surveyor can make the arrangements, but it should be part of the main contractor's responsibility to do so. Dealings with the contractor's domestic subcontractors should be only be through the main contractor with whose rates alone the surveyor is concerned.

Extent of measurement

The extent to which measurement will be necessary in making valuations for certificates will depend on the nature of the job and the stage it has reached. It may very often be possible to take the priced bill, identify the items that have been done, and to build up a figure in that way. Some items will, of course, be only partly done, and in that case a proportion will have to be allocated. At a first valuation, for instance, there may be little done beyond foundations, and to identify the appropriate items in the bill should not be difficult. If it is agreed that the foundations are two-thirds complete then the amount can be easily calculated. When, however, it comes to the superstructure, it may be necessary to take approximate measurements of such things as brickwork, floors and roofs. In some cases, to take measurements may involve excessive labour in relation to the value of the work, if it is practicable to take a proportion of a section of the bill: for example, half the plumbing bill or a quarter of the emulsion in the painting bill.

When dealing with housing, where a large number of similar units are involved, it should be possible from the bill of quantities to arrive at an approximate value of one house at various stages; for example

- brickwork up to damp-proof course
- brickwork to first floor level, with joists on
- brickwork to eaves
- roof on
- plastering and glazing complete
- doors hung
- plumbing and fittings complete
- decoration complete.

The value for different types of the same size of house will not vary sufficiently to make a difference for certificate purposes. The work done at any time can be valued by taking the number of houses that have reached each stage, and pricing out from the schedule, allowing in some cases, perhaps, for half and quarter stages.

On projects without firm bills of quantities the value of work carried out will be assessed through measurement of work on site or, on a straightforward project, by proportion of the individual items on the tender summary.

Towards the completion of the project, it is a good idea to check what is left to complete, as a safeguard against error made in a cumulative total. For the last two or three valuations the contract sum might be taken as a basis, and deduction made of (1) all PC or provisional sums and percentage additions and (2) work not yet done, the various accounts against (1) being added together with percentage additions *pro rata*, adjustment being also made for the approximate value of variations and price adjustment.

In all cases, throughout the contract the percentage addition or deduction on prices arising from insurances and/or rebate in the summary must be taken into account. It may be also found useful to have on file a note of the net amount of contractor's work (excluding provisional sums), so that an eye can be kept on the proportion of this included in each valuation. The value of certificates can be graphically represented to enable comparison with the anticipated cash flow. There may be reasons, such as site conditions or inclement weather, that result in a shortfall against that anticipated, but any serious departure from regularity should be looked into, as it may be an early indication of delays and difficulties being experienced by the contractor.

Preliminaries items

Each valuation will take into account the pricing of the preliminary bill or preliminaries items. These may be a number of items separately priced, or there may be one total for the whole. They may show in addition an allocation between fixed and time-related costs. Each priced item should be considered and a fair proportion of each included. Site establishment/removal costs and time related costs should be considered separately. The price for provision of offices, mess facilities and storage units could be split up into delivery cost, weekly rent and removal cost, and valuation made accordingly. An item for cleaning up on completion would not be included at all until the end of the job. If the items of site management are priced in the preliminaries, the sum shown could be divided by the period of the contract to give a suitable monthly sum. If it is anticipated that the contract time will be exceeded, suitable reductions should be made on the monthly figure to relate payments on account more accurately to the work actually carried out. If an extension of time award has been made this could affect the amount of preliminaries due.

Nominated subcontractors

In building up a valuation from a priced bill, all PC sums for nominated sub-contractors should be omitted and dealt with separately later (see specimen in Fig: 11.4). The subcontractors' claims will be taken one by one and examined, and a suitable figure added for each. The surveyor should have received either direct or from the architect a copy of the subcontractor's accepted estimate. This may or may not be the same as that used for arriving at the PC sum when the bill was prepared. It may give some detail of measurements and rates or may just give a lump sum. If the latter, and some subdivision is required, it can be asked for as a guide for valuations and a help at a later stage in measuring variations.

To arrive at an interim valuation of the subcontractor's work is sometimes very difficult. One may arrive on the site and find stacks of, say, metal windows and curtain walling in sections and pieces, with bags of bolts and fittings, and be presented with a statement 'To materials delivered: £5000'. There should be delivery notes indicating the portions of the subcontract for which materials have been delivered, and even though it is not practicable to make a complete detailed check it should be possible to assess the portion of the relative accepted estimate. Deduction would, of course, be made for the cost of assembling and fixing still to be done.

It is a good practice to notify each nominated subcontractor of the amount included in the surveyor's recommendation, though the architect is required by JCT 90, clause 35.13, to notify the subcontractor of the amount included on the certificate. The contract usually requires the surveyor to be satisfied that pay-ments included have been made, and this can be done either by asking the subcontractor in such a letter for advice when payments has been made, or by requiring the contractor to produce the receipts at the next valuation. Any contra charges showing up on the receipts should be reported to the architect. Named subcontractors under IFC 84 (see page 188) are treated for valuation purposes in the same way as the contractor's own domestic subcontractors.

Domestic subcontractors

The payment of interim amounts to domestic subcontractors is the responsibility of the contractor's surveyor. Applications for payment from subcontractors often form the basis of the main contractor's own application. Payments will be made to each subcontractor, generally in line with amounts certified to the main contractor. Reconciliations, on a monthly basis, will be required. The contractor's surveyor will analyse income received through certificate payments against costs of subcontractor payments and direct costs.

```
Job 123                    SOUTHTOWN SCHOOL
                                                              3/5/93
                           VALUATION NO 2
                                                    £           £
Bill No 1 Preliminaries

Start up                                          550
8 weeks at £980                                 7 840         8 390

Bill No 2 Groundworks            £

Site preparation complete                       1 497
Excavation complete            7 937
Less provisional items not done   600           7 337

Disposal complete             15 250
Less provisional items not done 1 700          13 550

Filling 90% complete                            8 300        30 684

Bill No 3 Concrete

Foundations complete                           19 349
RC columns 25% complete                         3 850        23 199

Bill No 4 Masonry

External walls 5% complete                                   1 300

Bill No 5 Drainage

Cast iron pipework 50% complete                              2 500

                                                            66 073
Add Profit and Overheads 4%                                  2 643

                                                            68 716

Nominated subcontractors

G&H Engineering   Off-site fabrication         1 500
        Add profit 2%                             30         1,530

                                                            70 246
Unfixed materials on site                                    1 300

                                                            71 546
                Less Retention 3%                            2 146

                                                            69 400
                Less Certificate No 1                       31 000

                Net Valuation No 2                         £38 400
```

Fig. 11.1 Sample valuation format.

Unfixed materials on site

Besides the value of work done, most forms of contract allow payment to be made to the contractor for unfixed materials brought to the site or, where the Standard Form is used, stored ready for the contract, though possibly subject to a higher rate of retention. The surveyor should ask for a list to be submitted by the valuation date, to be checked on the site visit. On substantial contracts the list can be certified beforehand by the clerk of works, who has the necessary data easily available. It is important, unless there is a special condition otherwise, that only materials actually on the site are included. The contract usually provides that unfixed materials so paid for shall become the property of the client, and if they are on the site they are under the control of the clerk of works.

JCT 80, clause 30.3, gives the architect discretion to allow the value of materials destined for the job and ready in all respects, but stored off the site, to be included in interim certificates, on condition that the provisos listed are complied with. A quantity surveyor faced with such an application should ask for the authority of the architect who, when satisfied as to content, labelling and other provisions of this clause, would in all probability give assent.

Problems have also arisen over retention of title of goods when certifying payments for materials on or off site. In the case of *Dawber Williamson Roofing Ltd* v. *Humberside County Council* (1979) 14 BLR 70 it was held that title in a quantity of roofing slates has not passed to the main contractor at the time he was paid by the client. The subcontractor was not paid, and the main contractor subsequently went into liquidation. As title had not been passed to the main contractor, the subcontractor (the plaintiff) was entitled to be reimbursed the value of the slates by the client. It is important therefore to ensure that such materials are, at the time they are paid for by the client, the lawful property of the contractor.

Price adjustment

When the traditional method of adjusting fluctuations is in operation, that is, by the use of clause 39 of JCT 80, it is valuable to start checking the records of price adjustment at an early stage – at any rate the labour portion – when a running total can be kept month by month for inclusion in the certificate valuation. Materials are rather more difficult to keep up to date than labour, owing to the time lag in rendering invoices. In either case increased costs, whether in respect of labour or of materials, should not be included in interim certificates unless supporting details have been made available.

Nowadays the more common method of adjusting fluctuations is by way of the price adjustment formulae (see page 185) where reimbursement of increased cost is automatic in each interim certificate by the application of the formulae to the current indices relevant to the proportions of the work categories actually carried out. Because these payments for fluctuations are final payments, the final account being simply a summary of the monthly totals, valuations will need to be more accurate than has been necessary in the past.

Retention

The amount to be retained must be checked from the conditions of contract as agreed. A special note should be made of any maximum retention fixed. In the

Standard Form a maximum of 5% of the contract sum is generally recommended (3% on contract sums over £500 000) but this may be found to vary. In the GC/Works/1 form (condition 48), for example, it is set at 5%.

When work is substantially complete, part of the retention sum is released, the balance being held as security for making good of defects that may be found necessary within the defects liability period. In the Standard Form half is to be released; in the GC/Works/1 contract this is also half, with an option to the authority to release further amounts at their discretion.

One of the changes that came about following the publication of JCT 80 was that each and every nominated subcontract is now treated as a small contract within the main contract with its own start and finish dates and its own certificates of practical completion. This means that nominated subcontractors are entitled to the release of part or all of the retention money held on their subcontract at varying times and no longer have to rely on the goodwill of the architect to certify release of their money before release becomes due to the main contractor.

On the whole question of release, the explanatory notes and the worked examples contained in the JCT guide to the Standard Form of Building Contract 1980 Edition will be found very helpful and worthy of study.

In the private edition of the Standard Form (clause 30.5.3) the retention money must be held by the employer as a fiduciary trustee in a separate bank account.

In some forms of contract the percentage retained of the value of unfixed materials is different from that retained from the value of work done, and the statement must be prepared accordingly.

Price adjustment is made without deduction of retention (JCT 80 clause 30.2.2.4) except when the price adjustment formula (clause 40) is used. Such adjustments have in recent years been largely increased, and being regarded as an out of pocket expense of contractors on which they receive no profit, it would not be reasonable to make the same deduction from them as is made from a value which includes a profit.

Predetermined stage payments

Predetermined payments on account, based on stage payments, have been traditionally used on large multiple contracts such as housing. To use such a method on a complex building contract is not feasible, but a variation has been introduced into the Third Edition of the Government contract GC/Works/1 whereby payments on account are based on predetermined payment charts. These charts, which are commercially obtainable, are based on extensive research into the pattern of payments on a variety of contracts. They take the form of a tabulated mathematical expression of a payment 'S' curve (the graphical expression of payment that traditionally represents the way money flows on a building contract: a slow start rising to a peak and then tailing off towards the end of the contract, producing an 'S' pattern).

A specimen stage payment chart is shown in Fig. 11.2. The contract period is one of the nine columns and the weeks of the contract period are shown down the right-hand side. As an example, in week 6 of a 55-week contract 1.98% of the basic sum is due. This basic sum is the contract sum less daywork provision (or contingencies if applicable), PC and provisional sums and provisional quantities.

Week	Proportion of basic sum payable								
	52	53	54	55	56	57	58	59	60
1	0.06	0.06	0.06	0.06	0.06	0.06	0.06	0.06	0.06
2	0.40	0.38	0.38	0.38	0.36	0.36	0.36	0.36	
3	0.61	0.59	0.57	0.57	0.55	0.55	0.55		
4	0.85	0.84	0.84	0.80	0.80	0.78			
5	1.20	1.20	1.18	1.18	1.16				
6	2.22	2.00	2.00	1.98					
7	4.23	4.00	4.00						
8	6.59	6.48							
51	99.91	96.90	90.87	82.50	79.87	76.90	71.33	69.90	66.45
52	100.00	99.89	94.75	90.06	86.78	81.58	79.99	74.56	74.00
53		100.00	99.89	95.64	90.01	89.98	84.64	80.90	81.65
54			100.00	99.65	96.21	95.11	90.87	88.21	85.66
55				100.00	99.54	97.00	94.11	90.89	90.01
56					100.00	99.00	97.00	95.61	93.22
57						100.00	99.00	97.32	96.83
58							100.00	99.00	97.75
59								100.00	98.83
60									100.00

Fig. 11.2 Stage payment chart.

To this basic sum the percentage showing from the chart is applied and to the resultant are added sums for nominated subcontractors, expended provisional sums, variations, daywork and fluctuations, if applicable. After deduction for retention on those parts of the total that are subject to retention, previous payments on account are deducted and the amount due emerges.

Facility is incorporated in the contract whereby if it is shown that the contractor is running early or late the contract administrator has the power to vary the payments by instructing the use of weeks earlier or later, as the case may be, than the actual week of the valuation.

Each tendering contractor is provided with a chart for the contract, but in the case of nominated subcontractors they are usually required to be prepared and submitted by the subcontractors as part of their tender.

Use of the charts allows both parties – contractors and clients – to know in advance the probable cash flow for the contract and to arrange their finances accordingly. From the quantity surveyor's point of view a great deal of time is saved: the time taken to prepare an interim valuation is, under this system, a matter of hours not days, which it often is under the traditional system. For contractors the same saving of time is available, although for their requirements it is often necessary to know in more detail the value of work done for their own internal purposes and for paying domestic subcontractors where the average-all-trades basis used in preparing the chart may not align with the requirements of their particular trade.

Predetermined payments are not at the moment featured in the JCT standard forms of contract, but this may well change in the near future.

Previous certificates

A careful check should be made to ensure that the figure shown as already certified is correct, as a slip here may make a serious error in the valuation. The architect should confirm the amount of the previous payment, or the figure may be referred to as 'previous valuations', the architect being asked to verify before certifying.

Specimen forms

A specimen valuation is set out on the standard RICS certificate valuation forms in Fig. 11.3. These forms are similar in design to the corresponding RIBA certificate forms in Fig. 11.4 and give all the information necessary to enable the architect to complete the certificate. The architect is required under JCT 80 to notify nominated subcontractors of amounts included and to call for proof of payment.

Both of these duties may be delegated to the quantity surveyor, whose duty it becomes to issue the notification and be satisfied as to proof of payment when the time comes to agree the next certificate valuation. In the event of failure to produce such proofs, the remedy is a direct payment by the client to be deducted from the next certificate, rather than refusal to issue another certificate until such time as the subcontractor is paid.

Nominated suppliers are not included in the list. They are merchants in exactly the same relationship to the contractor as other merchants, except that they have been selected by the architect, who has approved the price to be set against a PC sum.

Care should be taken not to refer to the valuation as a 'certificate' nor to the surveyor as 'certifying'. The surveyor only recommends, and it is for the architect to certify, who may take into account other matters than those within the surveyor's sphere, such as defective work for example.

When IFC 84 is used, because there are no nominated subcontractors, different standard forms for valuations and certificates have to be used, and these are available from the RICS and RIBA.

Certificate valuation papers

The quantity surveyor's copy of each statement should be kept together with all the papers relating to one valuation. The attached papers will include such things as subcontractors' applications, lists of unfixed materials, and interim statements of price adjustment.

Valuation on insolvency

In the event of the contractor's becoming insolvent it is prudent to make a valuation of the work executed, by taking the necessary measurements of the work up to the stage at which work ceases or is continued by another contractor. The valuation will include unfixed materials and plant, which the architect, or the clerk of works on the architect's behalf, will be responsible for seeing are not

Valuation

Quantity Surveyor
R.S.T. F/FRICS
Address

Works
Southtown School
New Extensions

Valuation No   2
Date of issue   Date
QS Reference   567

To Architect/S.O.

L.M.N., Esq., F.R.I.B.A.,
Address

Employer

Blankchester Diocescan Board of Finance,
Address

Contractor

U.V.W. (Contractors) Ltd.

Contract sum   £ 188,250. 00

© 1980 RICS

I/We have made, under the terms of the Contract, an Interim Valuation

as at   † and I/we report as follows –

	£	
Gross Valuation (excluding any work or material notified to me/us by the Architect/S.O. in writing, as not being in accordance with the Contract).	41,128	
Less Total Amount of Retention, as attached Statement.	1,188	
	£	39,940
Less total amount stated as due in Interim Certificates previously issued by the Architect/S.O. up to and including Interim Certificate No		28,000
Balance (in words) Eleven Thousand Nine Hundred and Forty Pounds	£	11,940

Signature   Quantity Surveyor

Notes:
(i)   All the above amounts are exclusive of V.A.T.
(ii)   The balance stated is subject to any statutory deductions which the Employer may be obliged to make under the provisions of the Finance (No. 2) Act 1975 where the Employer is classed as a "contractor" for the purposes of the Act.
(iii)   It is assumed that the Architect/S.O. will:-
     (a) satisfy himself that there is no further work or material which is not in accordance with the Contract.
     (b) notify Nominated Sub-contractors of payments directed for them and of Retention held by the Employer.
     (c) satisfy himself that previous payments directed for Nominated Sub-contractors have been discharged.
† (iv)   The Architect/S.O.'s Interim Certificate should be issued within seven days of the date indicated thus
(v)   Action by the Contractor should be taken on the basis of figures in, or attached to, the Architect/S.O.'s Interim Certificate.

Fig. 11.3 Valuation forms.

Statement of Retention and of Nominated Sub-Contractors Values

Quantity Surveyor
R.S. & T., F.F.R.I.C.S.

Works
Southtown School
New Extensions

This Statement relates to :-
Valuation No: 2
Date of issue   Date
QS Reference   567

	Gross Valuation	Amount subject to:			Amount of Retention	Net Valuation	Amount Previously Certified	Balance
		Full Retention of 3%	Half Retention of %	No Retention				
	£	£	£	£	£	£	£	£
Main Contractor	34,628	33,093	-	1,535	993	33,635	28,000	5,635
Nominated Sub-Contractors:-								
G & H Engineering	4,000	4,000	-	-	120	3,880	-	3,880
I & J Electrical	1,500	1,500	-	-	45	1,455	-	1,455
K & L Metal Windows	1,000	1,000	-	-	30	970	-	970
TOTAL	41,128	39,593	-	1,535	1,188	39,940	28,000	11,940

No account has been taken of any discounts for cash to which the Contractor may be entitled if discharging the balance within 17 days of the issue of the Architect/S.O.'s Certificate.
The sums stated are exclusive of V.A.T.

© 1980 RICS

Fig. 11.3 (cont.)

Statement of retention and of Nominated Sub-Contractors' values

Architect: L.M.N., Esq., F.R.I.B.A.,
address:

Works: Southtown School New Extensions
situated at:

Relating to Valuation No: 2
Job reference: 1234
Issue date: Date

| | Gross valuation | Amount subject to: | | | Amount of retention | Net valuation | Previously certified | Balance due |
		Full retention of 3%	Half retention of - %	Nil retention				
	£	£	£	£	£	£	£	£
Main Contractor:	34,628	33,093	-	1,535	993	33,635	28,000	6,535
Nominated Sub-Contractors:								
G & H Engineering	4,000	4,000	-	-	120	3,880	-	3,880
I & J Electrical	1,500	1,500	-	-	45	1,455	-	1,455
K & L Metal Windows	1,000	1,000	-	-	30	970	-	970
Total (The sums stated are exclusive of VAT)	41,128	39,593	-	1,535	1,188	39,940	28,000	11,940

No account has been taken of any discounts for cash to which the Contractor may be entitled if discharging the balance within 17 days of the issue of the Architect's Certificate.

This form is adapted from one of the same title published by the Royal Institution of Chartered Surveyors for their members' use. We are grateful to the RICS for releasing their copyright in this instance. RIBA Publications Ltd 1980

Fig. 11.4 Interim certificate forms.

Architect: L.M.N., Esq., F.R.I.B.A.
address:

Interim certificate
and Direction

Employer: Blankchester Diocesan Board of Finance   Job reference: 1234
address:

Interim Certificate No: 2

Contractor: U.V.W. (Contractors) Ltd.   *Issue date: Date
address:

Valuation date: Date

Works: Southtown School New Extensions
situated at:

* This date is to be not
later than 7 days from
the Valuation date.

Original to Employer

Under the terms of the Contract dated   date

in the sum of £ 188,250   for the Works named and situated as

stated above

I/We certify that the following interim payment is due from the Employer to
the Contractor; and

I/We direct the Contractor that the amounts of interim or final payments
to Nominated Sub-Contractors included in this Certificate and listed on the
attached *Statement of Retention and of Nominated Sub-Contractors' Values*
are due to be discharged to those named.

Gross valuation inclusive of the value of Works by Nominated

Sub-Contractors .... £ 41,128

Less Retention which may be retained by the Employer as detailed on

the Statement of Retention .... £ 1,188

£ 39,940

Less total amount stated as due in Interim Certificates previously

issued up to and including Interim Certificate No. 1 .... £ 28,000

Amount due for payment on this Certificate .... £ 11,940

(in words) Eleven Thousand Nine Hundred and Forty Pounds

All the above amounts are exclusive of VAT

Signed   Architect

Contractor's provisional assessment of total amounts included in above

certificate on which VAT will be chargeable £   -   @   %

This is not a Tax Invoice

Fig. 11.4 (cont.)

removed from the site (see JCT 80, clauses 16 and 27, GC/Works/1, conditions 56, 57 and 8). The purpose of such a valuation is that all parties may be aware of the financial position and some estimate of money outstanding to the insolvent contractor can be made.

Release of retention

Under JCT 80, clause 17.1, the architect is required to issue a certificate of practical completion of the works, on the issue of which the contractor is entitled under 30.4 to release of half the amount retained. Under GC/Works/1, condition 49, half the reserve is released on completion. The contractor will probably raise the question of release on one of the last interim certificate valuations, and the surveyor can suggest to the architect that the necessary stage is being reached. It is, of course, for the architect to decide.

When the architect has to release final balances to nominated subcontractors before that of the general contractor, a statement will be required giving the agreed totals of their accounts and showing the amounts of such final balances. If the variation account has not been forwarded to the architect, the subcontractors' accounts might be sent or, at any rate, any points which need the architect's confirmation should be raised.

Application by the general contractor for the release of the final balance of the retention on completion of maintenance work will probably be made direct to the architect, but it may be that, if the accounts are not completed, the need for this release will be applied as a spur to urge the surveyor to report the final figure.

Completion of contract by stages

Mention has been made earlier of the erection of buildings by stages. In such cases the contract should provide for a maximum retention for each stage; for certificate purposes each stage will be treated as if it were a separate contract. For housing estates where completion will be in small units of a few houses or flats, it may be that there will be no maximum retention less than the normal percentage, but provision will be made for prompt release of part retention as each unit is handed over, and of the final balance in similar stages.

Although not altogether relevant to valuations, note here that JCT 80 sets out how completion of a contract by stages shall be handled financially, and provision is made in the sectional completion supplement for this to be a contract condition. This was highlighted by a case (*M.J. Gleeson (Contractors) Ltd*. v. *London Borough of Hillingdon, Estates Gazette*, 8 July 1970) where the phased hand-over required was described in detail in the bill of quantities but not referred to in the conditions of contract and the client failed in a claim for damages for late completion of a phase.

Cost control and reporting

The subject of cost control is covered in Chapter 8; however, specific aspects relate to evaluating and reporting the cost implications of contract instructions and other issues affecting the out-turn cost of a project. Important as cost control is at the design stage, it is equally important during the progress of the contract to prevent either architect or client from ordering additional works without fully realizing their effect on the final account.

The client's quantity surveyor, if working closely with the members of the design team, should be aware at the earliest opportunity of proposed variations to the contract, including drawing amendments. Advance knowledge of proposed changes enables a full evaluation in terms of cost, quality and programming implications to be carried out in advance of their issue.

Although the initial estimate of variations to the contract is likely to be of a budgetary nature based on approximate measurements and notional rates or merely lump sums, it is important that such estimates be progressively updated as more detailed information becomes available in the form of firm measurements, quotations or daywork records. It is also necessary for the surveyor to review all correspondence and meeting minutes issued on the project in order to identify the potential cost implications of the issues contained therein.

Regular financial reports will be required to advise the client of the anticipated out-turn costs. As mentioned earlier the report will be tailored to meet specific client requirements. Certain clients will only require a simple summary statement of the current financial position (see example in Figure 11.5); others will require a detailed report identifying the cost implication of each instruction (whether issued or anticipated), which on a complex project may result in a lengthy document.

The regular report will identify adjustments to the contract sum in respect of the following:

- issued instructions
- adjustment of PC and provisional sums
- remeasurement of provisional work
- dayworks
- increased costs (if applicable)
- provision for claims and anticipated future changes.

An updated cash flow as referred to in Chapter 8 may also be included with the report to identify for the client the current level of expenditure against that anticipated in order for budgetary provisions to be adjusted accordingly. Supporting calculations should be filed with each report for future reference when it comes to reviewing and updating the costs.

Final accounts

This section deals with the principles of measuring for variation accounts and the practical implications of contract conditions covering the adjustment of quantities.

NEW C of E PRIMARY SCHOOL, SOUTHTOWN

for

BLANKCHESTER DIOCESAN BOARD OF FINANCE

FINANCIAL REPORT Nr. 6

L M & N	Date:	3.9.93
Chartered Architects	Contract Week:	27
	Date for possession:	1.3.93
	Date for completion:	27.5.94

		£
Approved contract sum:		726 547
Less Contingencies etc:		35 000
		£691 547

Estimates for instructions issued	£	
A.I's 1 – 11 as report Nr. 5	10 430	
A.I. 12 Ironmongery	586	
13 Ramp	850	
14 Redesign of CR Stores	(200)	11 666
		£703 213

Estimates for instructions to be issued	£	
Playground equipment	1 500	
Additional paving	600	2 100
		£705 313

Ascertained claims	—
Estimated final total (excluding contingencies)	£705 313
Current approved sum	726 547
Balance of contingencies	£21 234

Notes

3 Weeks extension of time granted. Completion date now 17.6.94 anticipated completion date 15.7.94 i.e. Contractor estimated to be in 4 weeks of culpable delay.

Loss and expense claim received in the amount of £26 500 not ascertained or certified – not included above.

All figures exclusive of V.A.T. and professional fees.

Fig. 11.5   Financial statement.

Architect's instructions

Once a contract has been concluded its terms cannot be changed, unless the contract itself contains provision for variation, or the parties make a further valid agreement for alteration. The standard-form contracts contain extensive machinery for variation, but the only variations thereby permitted are those that fall clearly within the contractual terms. If the desired change is not covered by those terms it can only be effected properly by fresh agreement. In this connection care should be exercised to ensure that the new agreement is itself a valid contract and, in particular, that it is supported by consideration given by both parties.

When making visits for interim certificate valuations the surveyor should keep an eye on variations in the contract that have arisen. It is often valuable to have seen the work in course of construction. The responsibility for issuing instructions is with the architect, but the surveyor requires these as an authority to measure. Provisional sums, PC prices and provisional quantities can be automatically adjusted without any instruction. The contract may also provide (JCT 80, clause 2.2; GC/Works/1, condition 3) that errors in the bill of quantities shall be treated as a variation and adjusted. No specific instruction will then be required for doing this, but the architect who has to certify the final account should be told of any substantial items.

The surveyor will probably have received revised drawings from the architect from time to time. These should be dated on receipt, and it might be found advantageous to mark them with a large V (variation drawing) in the bottom right-hand corner. This will distinguish them clearly, when referred to later, from the drawings on which the tender documentation was based.

A separate file for the formal architect's instructions should be kept with the variation papers. Besides these, the surveyor will receive copies of correspondence that affects variations. Such letters may be found to be explanatory of the orders and indicate their intentions, so helping when it comes to measurement.

Site meetings

On most contracts the architect will probably arrange for meetings on the site at regular intervals – perhaps monthly or fortnightly – of those concerned in carrying out the contract, including consultants, clerk of works, contractor and subcontractors. Quite often the quantity surveyor is invited to attend such meetings, but as they are principally concerned with settling details and ensuring progress, such presence is apt to be superfluous. If estimates are required, there is little that can be done during the meeting, except in a very general way. However, as a means of keeping in touch with the circumstances under which variations arise, attendance at such meetings is useful, if the time is available and should the fee level allow.

Proposal to start measuring variations

At a certain stage, in order not to leave too much work to the end of the contract, the surveyor will probably want to start the measurement of variations. Formal architect's instructions may not be available, but a fairly good idea as to what the variations are from knowledge of the bill of quantities and the work in progress can be obtained. Before beginning measurement, the contractor must be advised in writing that it is proposed to do so (specifically required by JCT 80 clause 13.6; GC/Works/1, condition 37) and an appointment should be made with the contractor's surveyor. Most large firms of contractors will have their own staff of surveyors, but it is not uncommon, particularly on small contracts, for the contractor to leave the preparation of the account entirely to the quantity surveyor, being mainly interested in the final total, and not considering it worth the expense of going through all the detail.

It is not advisable to start too soon on measurement of variations when future developments, which cannot be foreseen, might affect the surveyor's work. One might, for instance, measure a number of adjustments of foundations or drains, only to find later that the whole of one of these sections must be remeasured complete. However, when visiting a distant site for an interim valuation, which only occupies part of the day, it is useful to arrange with the contractor to spend any time available on work for the variation account.

It is fatal to postpone the measurement of variations for too long. The benefit of fresh memories is lost and the accumulation of such work may be a serious strain at a time when there is pressure in other directions. Remember that contractors are entitled to prompt payment for work done in such instalments as the contract provides, and both extra works and deductions should be assessed without delay to avoid penalizing them. Contractors have told stories about accounts years late which, if the delay was the surveyor's fault, are a serious blot on the surveyor's character.

Hidden work

If there is a clerk of works the quantity surveyor should ensure that arrangements are made for records of hidden work to be kept in the form required. Clerks of works vary, of course, in their ability and experience, and it would be unwise to assume that the clerk of works knows exactly what is required without any guidance. The depths of foundations, position of steps in foundation bottoms, thickness of hardcore and special fittings in drainage are all examples of items that the clerk of works may be asked to note and record. If these records are carefully kept and agreed at the time with the foreman, the quantity surveyor and contractor's surveyor should have no difficulties from lack of knowledge.

Procedure

When starting to measure, the surveyor should either have the architect's instructions or a list of items of variation compiled for which variation orders will be requested. Each should be taken in turn and the relative measurements made. As a general principle, adjustment will be made by measuring the item as built and omitting the corresponding measurements from the original dimensions.

There will be occasions when it may be easier to adjust a contract item by either 'Add' or 'Omit' only. If all emulsion on walls is altered to eggshell paint, an 'Add' item of the contract quantity as extra cost only would be suitable.

It is very important to keep omissions and additions distinct, and it is suggested that the words 'Omit' or 'Add' should always be written at the top of every page and at every change from omission to addition. Each item of variation should be headed with a brief description and instruction number, if known. In accounts for public authorities, the architect's instruction references may well be required by the auditors.

It is not essential that the omissions shall be set down in the dimension book at the same time as the additions are measured. It is more usual while on site to measure the additions, leaving the omissions to be looked up and put down in the office. Sometimes complete items of additions can be measured from detail drawings in the office, particularly if the contractor has no surveyor taking down all the measurements. If a surveyor acts for the contractor, it may be necessary to visit the quantity surveyor's office to agree measurements. However, the contractor's surveyors may be satisfied to let the quantity surveyor do this alone, raising any points on measurement after having examined the account. The quantity surveyor must be prepared to produce the original dimensions to the contractor's surveyor to verify measurements taken from them. but it is unreasonable to be expected to do the work twice, and a reasonable arrangement should be made with the builder's surveyor in such a situation.

The surveyor should be supplied with a copy of specialists' estimates accepted by the architect, so that full particulars are available from which to check their accounts. Many such estimates are subject to measurement as executed, and if no copy is supplied to the surveyor, the architect's will have to be borrowed for reference when measuring.

See Fig. 11.6. For a sample final account entry.

Grouping of items

Before any abstracting or bill of measurements is started the surveyor should decide on the suitable subdivision into terms which will be adopted in the account. Since items may or may not correspond with variation orders, they may be arranged in a different order: a variation order may be subdivided or several grouped together, if their subject-matter suits.

Quite possibly the adjustment of foundations will be the first variation for which measurements are taken. At this stage it may not be known what other variations there will be, but the foundation adjustment can be regarded as an item with which other variations will not interfere. Supposing there is a minor change in plan for which a variation order is issued: that change will affect foundations, so that there may be a variation within a variation. Unless there is any special reason for distinguishing, the lesser variation will be absorbed in the greater. When it comes to adjusting for the change in plan, this will be done for the superstructure only.

It might, however, happen that the complete value of the change in plan is separately required. For example, in the rebuilding of a fire-damaged building it might be that a change in plan was being made at the client's request and expense. In that case the foundation adjustment would have to be subdivided to

ITEM 8 (Continued)
DRAINAGE
OMISSIONS (Continued)

£

A.I. 6.19, 7.3 and 8.7

Bq bf

15/2A   Excavate trench average 250 mm deep   15   m   1.52   22 | 80

15/4C   100 mm Granular bed to 100 mm pipe   15   m   3.44   51 | 60

Etc   Etc

15/6A-8B   Nr 8 Manholes   Complete   1,986

Etc   Etc

·To Collection   £

ITEM 8 (Continued)
DRAINAGE
ADDITIONS (Continued)

£

A.I. 6.19, 7.3 and 8.7

15/2A   Excavate trench average 250 mm deep   3   m   1.52   4 | 56

*   Ditto average 500 mm deep   20   m   2.77   55 | 40

Etc   Etc

Nr 10 Manholes

15/2C   Excavate pit not exceeding 2.00 m deep   15   m³   12.37   185 | 55

15/3A   Remove surplus from site   15   m³   6.05   90 | 75

Etc   Etc

11/25

To Collection   £

Fig. 11.6   Sample final account entry.

give the separate costs required. If the variation is a completely additional room, then the foundation measurements for that room can easily be kept separate from those for the foundations generally, which is preferable as it gives more accurate relative values.

As the list of variations develops the surveyor will be able to decide on how to group them. For instance, there may be one order for increasing the size of storage tanks, another for omission of a drinking-water point and a third for addition of three lavatory basins. Each of these will be measured as a separate item, but the surveyor may for convenience decide to group these together as 'variations on plumbing'. But if the client has ordered the three lavatory basins and does not know about the storage tanks or drinking-water point (which are changes in the architect's ideas), it may be advisable to have the value of the extra for basins separately available for reporting.

It will be convenient to group the very small items together under the heading of 'Sundries', preferably in such a way that the value of each can be traced.

Provisional quantities

It often happens that such work as cutting away and making good after engineers is covered by provisional quantities of such things as holes through walls and floors, or making good of plaster and floor finishings. The original bill may have been taken from a schedule supplied by the engineers, and the need for remeasurement on the site must not be overlooked. It does sometimes happen that the provisional quantities reasonably represent the work carried out and can therefore be left without adjustment, but this should not be done merely to avoid what is certainly a rather laborious job. Non-technical auditors are apt to frown on such procedure. One of the few things they can do to check a technical account is to go through the original bill and see that all provisional items have been dealt with. An appendix to the variation account, showing how this has been done, can be of help to an auditor.

Daywork

Certain variations, which it may not be reasonable or possible within the terms of the contract to value at contract rates or rates analogous thereto, may be charged on a prime cost basis. Daywork sheets will be rendered for these by the contractor, which set out the hours of labour of each operative and a list of materials used. If there is a clerk of works it will be one of the duties to certify that the time and material are correct. The clerk of works' signature is not in any way authority for a variation, nor does it signify that the item is to be valued on a daywork basis instead of by measurement. When there is no clerk of works the architect will generally sign the sheets. Neither architect nor surveyor, if they are not continuously on the site, can directly guarantee that the time and material are correct, but, if these appear unreasonable for the work involved, they can make enquiry to satisfy themselves.

JCT 80 clause 13.5.4 provides for pricing daywork as a percentage addition on the prime cost.

The definition of prime cost is laid down in *Definition of Prime Cost of Daywork Carried Out Under a Building Contract* (RICS), and the percentages required by the contractor are to be filled in in a space to be provided in the contract documents.

For work within the province of some specialist trades, (electrical and heating and ventilating), there are different definitions of prime cost agreed, and these must be taken into account in the preparation of subcontracts. These have been agreed between the RICS and the Electrical Contractors' Association and the Electrical Contractors' Association of Scotland, and between the RICS and the Heating and Ventilating Contractors' Association.

The provision made for daywork should be taken into account when considering the amount of any provisional sum for contingencies.

Overtime

Though there is nothing to prevent a contractor's employee working overtime (subject to trade union control), this is normally entirely a matter for the contractor's organization. No extra cost of overtime can be charged without a specific order. Where, therefore, overtime is charged on a daywork sheet, it will be entered at the standard time rates, unless there is some such special order.

It may be that, owing to the urgency of the job, a general order is given for overtime to be worked, the extra cost to be charged as an extra on the contract. Or the order may be a limited one with the object of expediting some particular piece of work. When an operative paid, say, £10.00 per hour works an hour a day extra at time-and-a-quarter rate, i.e. £12.50, the ¼ hour (£2.50) will be chargeable in such cases. As a matter of convenience on the pay-sheet, if the normal day is 8 hours, the entry, instead of being 8 hours @ £10.00 and 1 hour @ £12.50, will be 9¼ hours @ £10.00. The ¼ hour is not 'working time' at all, and is therefore sometimes called 'non-productive overtime': the extra cost of payment for overtime work over normal payment. Any charges that are to be based on working time, such as daywork (where overtime is not chargeable as extra), must exclude the ¼ hour. Where the extra cost of overtime is chargeable, the data will be collected from the contractor's pay sheets and verified if necessary from the individual operative's time sheets.

Price adjustment

Labour
The traditional way of adjusting fluctuations in the cost of labour and materials is by way of a price adjustment clause (JCT 80, clause 39). Under such a clause any fluctuation in the officially agreed rates of wages or variation in the market price of materials is adjusted. JCT 80 (clauses 38.7 and 39.8) provides for a percentage to be inserted in the appendix to the contract, at the discretion of the employer, to be applied to all fluctuations to allow for contractor's profit and overheads.

Even with such a clause the contractor can still be faced with unexpected expenses, such as occurred some years ago when the increase of insurance contributions arose on the establishment of the National Insurance scheme (neither anticipated nor covered by the Standard Form of the time). This was neither payment of wages nor cost of materials and in most cases was therefore not recoverable, except as an *ex gratia* allowance made by the client. JCT 80 makes provision for such sums to be recovered by clause 38 (in which case 39 is not used).

The checking of wages adjustment should be fairly straightforward on an examination of the contractor's pay-sheets. The rates of wages are officially published by the National Joint Council for the Building Industry (NJCBI), so there should be no doubt as to the proper amount of increases or decreases or the dates on which they came into effect. To these increases will be added allowances for increases arising from any incentive scheme and/or any productivity agreement and for holiday payments as set out in JCT 80 (clause 39.1.1). These increases will apply to workpeople (defined in clause 38.6 and 39.7) both on and off the site and to persons employed on site other than workpeople (clause 39.1).

Care must be taken that there is no overlapping with the rates charged for daywork when dealing with price variations. If daywork has been priced at actual rates (as required by JCT 80, clause 13.5), say with labour 5p per hour above basic rates, the number of hours so charged in daywork must be deducted from the total on which price adjustment is being made. In this way the contractor gets, for the hours charged in daywork, a percentage on the difference in cost, whereas adjustments under the price variation clause are strictly net differences. If, of course, the contract provides for daywork to be valued at basic rates, the point does not arise.

GC/Works/1, condition 42 (4)(d), requires the rates for valuing dayworks to be provided for in the bill of quantities.

Materials
The adjustment of materials prices is more difficult. The contractor will produce invoices for those materials from which the quantities and costs can be abstracted and the value will be set against the value of corresponding quantities at the basic prices. Prices must be strictly comparable. If the basic rate for eaves gutters is for 2 m lengths, an invoice for 1 m lengths cannot be set against that rate. The 1 m length rate corresponding to the 2 m length basic rate must be ascertained. There is also the difficulty of materials bought in small quantities, perhaps by the foreman from the local ironmonger, when again the price paid is not comparable with the basic rate. JCT 80, clause 39, says 'if the market price ... increases or decreases' and these are material words. When in doubt the applicability of the contract wording must be considered.

As for labour, reference must be made to the rates charged in daywork for materials and adjustment made, if necessary, on the totals being dealt with for price adjustment.

Invoices should be called for in respect of *all* materials appearing in the basic list, as the surveyor is responsible for seeing that fluctuations in either direction are adjusted. This is another case where non-technical auditors are apt to worry if all items do not appear in the account.

The surveyor should also see that the quantities of the main materials on which price adjustment is made bear a reasonable relation to the corresponding items in the bill of quantities and variation account. An approximation, for instance, can be made of the amount of cement required for the concrete and brickwork, and any serious discrepancy should be investigated.

Formula method
The more common method of price adjustment in building contracts nowadays is by way of formulae to calculate the adjustment.

Two formula methods are used in the UK, one for building contracts and the other for civil engineering work. They both use the same formula for calculating increased costs, but each uses different work categories, which relate to the type of work to be performed. They employ the use of cost index numbers, which are recorded on a national basis. The formula then utilizes these indices to measure the average changes in construction costs.

The concept of these formulae is different from that of the traditional reimbursement provisions for price fluctuations. These latter are said to leave a considerable shortfall in overall recovery on the contract and, what is worse, a shortfall that is unpredictable, particularly in unstable economic circumstances. Formulae methods greatly simplify the administration of price fluctuation provisions, facilitate prompt payment of fluctuations on interim valuations, and reduce the scope for dispute. Contractors can quote competitively on current prices with the confidence that reimbursement will be in terms of current prices throughout the contract.

Two documents (*Price Adjustment Formulae for Building Contracts (Series 2): Guide to Application and Procedure* and *Price Adjustment Formulae for Building Contracts (Series 2 Revised): Description of the Indices*, HMSO) have been published that explain the formulae and provide information and assistance to those using them. The formulae are of two kinds, the building formula and specialist engineering installations formulae.

The building formula uses standard composite indices (each covering labour, materials and plant) for similar or associated items of work which have been grouped into 49 work categories. The first eight work categories are given by way of example:

(1)   demolitions
(2)   site preparation, excavation and disposal
(3)   hardcore and imported filling
(4)   general piling
(5)   sheet steel piling
(6)   concrete
(7)   reinforcement
(8)   structural, precast, and prestressed concrete units.

The formula is applied to each valuation, which will need to be separated into the appropriate work categories. The formula method cannot be applied to any approximate valuations made between the usual monthly certificates.

There are alternative applications of the formula available. Each of the 49 work category indices may be applied separately. This provides the most sensitive possible application of the formula. Alternatively, the 49 work categories may be grouped together to form work groups. Clearly the fewer work groups used, the less sensitive will be the indices to changes. It must also be practicable to analyse the tender and the value of work carried out in each valuation period into the selected work groups. This entails less work in separating the value of work carried out in every valuation. This application of the formula to the main building contract does not prevent the use of one or more of the 49 work categories to subcontracted work should the parties so desire.

These alternative uses are described in detail in the *Guide*, which also gives notes on the application of the formula at precontract, interim valuation and final account stages, with sample forms and worked examples.

The specialist engineering installations formulae cover electrical installations, heating, ventilating and air-conditioning installations, lift installations, and structural steelwork installations. They are applicable whether the work is performed by direct contract or by nominated subcontract. These formulae use separate standard indices for labour and for materials, the respective weightings of which are to be given in the tender documents, except for lift installations where the weightings are standardized. In each case the formula is expressed in algebraic terms and has been devised in conjunction with the appropriate trade association. It is intended that these specialist formulae will normally be applied to valuations at monthly intervals.

Nominated subcontractors

The quantity surveyor is responsible for the checking of the accounts of nominated subcontractors, and when such accounts contain measurable items measurements will need to be taken, usually from the site, to check them. It is more satisfactory to meet and measure with the subcontractor than to wait for the account to be rendered. If measurements are agreed and taken together, there should be nothing factually wrong with the account when it comes in. And much argument over measurements and correspondence over credit notes may be saved.

The quantity surveyor should have received from the architect copies of the accepted estimates of nominated subcontractors, and these must be studied to see that the relative accounts are in accordance with them. They may include lists of basic prices of materials or merely state in general terms that the estimate is subject to adjustment in cost of labour and materials. If the latter, the subcontractor will have to be asked to submit a statement with, for materials, supporting vouchers. Labour rates will probably be governed by the working rules of the particular trade, and can easily be substantiated.

If any extra items are chargeable on a daywork basis, the rates should be fixed in the same way as provided for the main contractor. Under JCT 80 (clause 13.5.4) provision is made for the definition of prime cost of a particular trade association to be used, when works of a specialist nature fall within the province of such an association.

In checking price adjustment of materials, the discounts on invoices should be watched to ensure that basic rates and invoice rates are comparable from this angle.

As well as adjusting the nominated subcontractor's account against the PC sums it is also necessary to adjust the main contractor's profit and attendance. The profit, being cost-related, is usually priced in the bill of quantities as a percentage of the PC sum, and the same percentage is therefore applied to the subcontractor's total in the final account. Attendance, whether it be general or other attendance, is work-related. For example, unless there are some very special requirements, the attendance on a carpet-layer laying carpet at £5.00 per square metre is the same even if the carpet costs £20.00 per square metre.

Occasionally a contractor will price attendance as a percentage for convenience but more often it will be priced as a sum. Consequently, unless the amount of contractor's work that is defined as attendance has changed, then the attendance sum will not be adjusted and the amount included in the bill of quantities will be the same figure as in the final account. The fact that the original pricing was by way of a percentage will make no difference, and the sum will remain the same.

Named subcontractors

Named subcontractors are, as their title suggests, named by the architect and in this respect only are similar to nominated subcontractors. Once a subcontractor has been named (JCT 80, clause 19; IFC 84, clause 3.3) they become domestic subcontractors in exactly the same way as subcontractors chosen by the main contractor. No special requirements therefore arise regarding the quantity surveyor's duties to agree their final accounts; this is the responsibility of the main contractor and, unless variations have arisen, no special entries in a final account will be made.

Domestic subcontractors

The settlement of domestic subcontractors' final accounts is the sole responsibility of the contractor. The principles for valuation of variations, measurement and daywork procedures will be similar to those under the main contract. Each individual subcontract final accounts will be negotiated separately, often based on build-ups submitted by the subcontractor. It is common for subcontract conditions to make provision for 'set-off' by the main contractor between one subcontractor and another.

Precosted variations

The preparation and agreement of a final account can be a time-consuming and expensive process for both parties, so much so that in many cases it has been found that the measurement period often exceeds the contract period. In an attempt to overcome these problems the principle of precosting variations has been introduced into JCT 80 by way of a new clause 13A and in the new Government form of contract GC/Works/1.

Under these conditions of the contract the contract administrator has the power, if thought fit, to seek a firm price quotation from the contractor for variations before confirming the order. Notification of the intended variation is sent to the contractor, who is then required to price the instruction, breaking the cost down into the cost of the work and the cost of any concomitant prolongation and disruption. The price is then submitted for scrutiny by the quantity surveyor, who passes recommendations to the project manager who, if the price is acceptable, confirms the instruction. This procedure is subject to a strict timetable, and provision is made for adequate information to be supplied by both the contract administrator and the contractor. If an acceptable price cannot be achieved using this procedure then traditional methods of agreeing the cost are adopted.

Under the terms of clause 13A of JCT 80 this procedure is subject to the agreement of the contractor who is required, in addition, to provide a method statement. Under the Government form the decision is solely that of the project manager.

Precosting of variations will obviously save a great deal of time and go a long way to alleviating the problems referred to above. There are drawbacks however, not least of which is that precosted variations tend to be more expensive than those arrived at by traditional means. This is because, while the pricing of the work itself may not cause too many problems, the costing of prolongation and disturbance (both totally unknown factors at the time of pricing) has to be something of a gamble, and contractors must err on the pessimistic side in order to safeguard their position. This is, however, often acceptable to clients; while they may have to pay more, at least they have a firm figure rather than an estimate from a quantity surveyor, which will not be confirmed for many months and may, often by no fault of the quantity surveyor, turn out to have been unreliable.

Bill of variations

When the items are billed they should be priced out. All contract rates that are binding should be written in ink and the surveyor's analogous or new rates in pencil. The contractor can then see at once which rates are subject to negotiation. It is helpful if the item numbers in the bill from which the analogous rates are built up are marked against the rates to show the contractor their basis. Note that contract rates are only binding if applied to work under similar conditions (see JCT 80, clause 13.5.1). Any very serious difference in quantity, for instance, would justify a varied rate. A bill containing 500 m of 150×25 mm moulded wood skirting might be so varied that only 5 m are required. It is obvious that this length cannot be made to detail at the same rate per metre as 500 m but, generally speaking, such departure from contract rates will only be made in extreme cases. A certain amount of 'swings and roundabouts' must be expected.

It will probably be found that a number of invoices are required from the contractor for costs to be set against provisional sums and PC prices. It is best to make a list and send it to the contractor, who then knows exactly what to look out for. These will be examined, compared with estimates and entered in their place in the account, profit being added *pro rata* with that in the contract, and any adjustment being made on the attendance item. If this later is expressed as a percentage, it would also be adjusted *pro rata*.

When the account has been priced out, there will probably be some blanks that cannot be completed. It may be that further information on some point is required from the contractor. The summary will be made as far as possible and approximate figures added for the blanks to give the surveyor some idea of how the account is coming out. A question from architect or client on this subject may be expected at any time or an application for more money from the contractor, and it will then be possible to give an approximate figure. In reporting to architect or client some allowance should be made for possible adjustment of prices after examination of the account by the contractor, to be on the safe side. A copy of the draft variation account will then be forwarded to the contractor; see example format of a summary and statement of final account in Fig. 11.7.

Meeting with contractor

The contractor, having examined the account, is fairly certain to have some criticism. Unless the criticisms are of a minor nature, which can be settled by correspondence, an appointment will be arranged for the contractor's surveyor to call at the quantity surveyor's office and go through the points.

There may be differences in the quantities between the surveyor's bill and the

Job 123

New C of E PRIMARY SCHOOL, SOUTHTOWN

SUMMARY OF VARIATION ACCOUNT

ITEM NO		OMISSIONS £	ADDITIONS £
1.	Preliminaries	—	1 600.00
2.	Substructure	1 328.61	2 413.60
3.	Reinforced concrete construction	4 284.31	8 029.41
4.	Roof coverings	127.42	1 886.19
5.	Windows	1 241.91	2 561.24
6.	Floor finishes	—	1 661.41
7.	Sanitary fittings	496.00	891.60
8.	Drainage	1 094.63	3 189.20
9.	Fittings	491.42	2 941.30
10.	Reduced length of centre wing	1 362.40	427.86
11.	Garden store	—	784.20
12.	Entrance road	699.41	2 213.90
13.	Playground equipment	—	4 194.20
14.	Sundry variations	—	7 148.62
15.	PC, provisional sums and contingencies	175 000.00	143 507.02
16.	Dayworks	—	7 394.20
		£186 126.11	£190 843.95

Add main contractor's profit and overheads

4% on total less Provisional sums (20,000)		6 645.04	—
4% on total less daywork (7,394.20)		—	7 337.99
Loss and expense settlement		—	8 950.00
		£192 771.15	£207 131.94
			192 771.15
Net addition carried to statement			£14 360.79

STATEMENT OF FINAL ACCOUNT

	£
Amount of contract sum	726 546.97
Net addition as summary	14 366.79
Amount of final account	£740 907.76

We confirm our agreement to the final account in the total sum of £740,907.76 and certify that we have no further claims under the terms of the contract.

Signed: _____ CONTRACTOR _____
                (Director)

Dated: _____

Fig. 11.7 Summary/statement of final account.

contractor's record as taken on the site with the quantity surveyor. Each will have to look up the original dimensions, and it may be found that something has been omitted or a mathematical or copying mistake made. A recollection of the item and the circumstances will probably put the matter right, and, if necessary, the surveyor will amend the account.

Differences in the pricing are more difficult to straighten out. Each will produce an analysis of the disputed price and the value must be argued. The contractor may well produce information as to the circumstances in which an item was carried out, of which the surveyor was unaware or did not fully realize, and may therefore feel justified in amending the price. With a little readiness to give and take on both sides there should be no difficulty in getting through the account satisfactorily.

Commercial pressures are sometimes brought to bear on the settling of final accounts, particularly when the client is keen for an agreed sum to be fixed as soon as possible. This may involve overall negotiation on the value of specific sections of the project without a detailed evaluation of each variation.

Completing the account

Once necessary alterations to the quantities and rates have been made as a result of the meeting, the extensions and costs will be corrected and the total agreed with the contractor. A copy will then be completed in ink or typed; and it will be found a useful practice to have all mathematics rechecked, to ensure that the actual document submitted is free from mathematical errors. If a clerical error is found the matter should be taken up again with the contractor and agreement obtained to the revision.

The percentage adjustment for insurances in the summary must not be forgotten. Preliminaries are not normally adjustable, but if the amount of the contract is very seriously increased, by the addition of another wing for example, the preliminary items should be adjusted *pro rata*, because the substantial extension of the contract time will involve increased expense of the agent, and use of plant, huts, telephone and other items that are priced in the preliminary bill.

A careful check is necessary to ensure that all PC and provisional sums have been adjusted. It is a useful practice to run through all such sums in the priced bill, marking them with a reference to the relative item in the variation account. Any that have not been dealt with will then show up clearly. When all provisional sums are together in an early bill, such a check is simplified. Special care is needed when some appear, say, in an external works bill: it has been known for such an item to be missed and the error even to pass a first technical audit. One of the things that a non-technical auditor will certainly do is to check that all such sums have been adjusted.

It is usual to ask the contractor to sign a copy of the final account as evidence of agreement, or to confirm the final figure in a letter. It may be advisable to see the architect with the account before final agreement with the contractor, in case something unexpected arises, or else to agree it 'subject to the architect's confirmation'.

For public authorities the full variation account and summary will usually be required and examined by the appropriate officer who understands the technicalities involved. However, for a private client, whether an individual or a board of directors, a simplified statement showing the principal variations and their value will probably be more lucid. The full account would need a good deal of explanation of its form, the arrangement of omissions and additions, and the meaning of provisional sums and PC items. This statement would be forwarded through the architect, with whom it may be advisable to discuss its form beforehand, and should be accompanied by the complete account for the architect's information, or for production to the client if demanded. The surveyor's fees if a percentage will be based on the value of the complete account. The client is, of course, entitled to have the complete account if required.

Audit

The variation account on a building contract will, for public authorities, always be subject to audit. After examination by the technical officers of the authority, the account will be scrutinized by the finance division (for a government department) or the district auditor of the Board of Inland Revenue (for a local authority).

Though, for a private client, the account may not receive any further financial check, the quantity surveyor must be prepared to have such a check made. A public company will have an accountant on its staff responsible for finance, and the company's auditors may raise questions. Even an individual client may feel dissatisfied and refer the account to an accountant.

Ultimately, questions might be asked in Parliament for public authorities, at a shareholders' meeting for a company, or in seclusion by a private client.

Contracts without quantities

For contracts without quantities a similar procedure will be followed. The bill of quantities will be replaced by a copy of the contractor's 'schedule of rates', which is required under the conditions of the relevant standard contract. This schedule will often be found to be in a much abbreviated form compared with a surveyor's bill of quantities, but must be used as a basis, so far as it goes. Remember that it is solely a schedule of rates, and that any quantities are not part of the contract. The quantity surveyor should see at the first opportunity after the contract is placed that this schedule is received. The scantier the schedule is, the more will the surveyor have to fall back on the 'fair valuation' rule.

Claims

Claims, in the shape of assertions by contractors that they are entitled to additional payment, are an ever-present feature of the construction industry. Once the contract is made, the contractor is responsible for carrying out precisely the required work, in the prescribed time, for the agreed sum. Generally an entitlement to any relaxation of those terms can only be acquired if some event occurs for which the contract itself provides for compensation, or if a breach of contract is committed by the client or a party for whom the client is responsible.

On most other occasions the costs lie where they fall and the contractor will have no recourse to recover them. Thus loss and delay arising from the intervention of outside parties unconnected with the contract almost invariably fall on the contractor. Furthermore, there is no privity of contract between employer and subcontractor, unless a supplementary contract is concluded between them. The fact that loss is sustained, without fault on the part of the loser, may merit sympathy but does not of itself demand compensation.

Where the standard forms of contract are in use, many claims are simply attempts to invoke the compensatory provisions of the contract or, from the contractor's viewpoint, to obtain entitlement thereunder.

The details of claims should be investigated by the surveyor if requested to do so by the architect and a report made to the architect/client. The report should summarize the arguments, which can be elaborated verbally, and set out the financial effect of each claim.

The quantity surveyor may well be requested to negotiate with the contractor on specific issues relating to a claim before there is resort to litigation. Particular care is required in the conduct of these negotiations as they may affect the outcome of subsequent legal proceedings. To guard against commitment, correspondence in the period before action is often marked 'without prejudice'. This will usually preclude reference to such correspondence in subsequent litigation but it is not, as is sometimes mistakenly believed, a safeguard against any form of binding obligation. In fact should an offer made in such correspondence be accepted, a binding agreement will usually result. This may be so even if the matter agreed relates only to a detail and the negotiations as a whole eventually collapse.

Though the decision on claims rests with the architect, except in such matters of valuation as the parties to the contract have entrusted to the surveyor, the quantity surveyor in a preliminary consideration of them should remember the principles that must guide the architect in a decision. The following thoughts are suggested as a guide to a decision on claims:

- What did the parties contemplate on the point at the time of signing the contract? If there is specific reference to it, what does it mean?

- Can any wording of the contract, though not specifically mentioning it, be *reasonably* applied to the point? In other words, if the parties had known of the point at the time of signing the contract, would they have reckoned that it was fairly covered by the wording?

- If the parties did not contemplate the particular matter, what would they have agreed if they had?

- If the claim is based on the contract, does it so alter it as to make its scope and nature different from what was contemplated by the parties signing it? Or is it such an extension of the contract as would be beyond the contemplation of the parties at the time of signing it? In either case the question arises whether the matter should not be treated as a separate contract, and a fair valuation made irrespective of any contract conditions.

- The value of the claim in money should not affect a decision on the principle.

If the claim is very small, however, whichever party is concerned might be persuaded to waive it, or it may be eliminated by a little 'give and take'.

In particular, any action of the client that may have been a contributory cause should be given due weight. If the claim is based on unanticipated misfortune, consideration of what the parties would have done, if they had anticipated the possibility, will often indicate whether it would be reasonable to ask the client to meet the claim to a greater or lesser extent.

If the architect is not able to give a decision, the matter will be referred to the client with a recommendation. If the contractor does not accept such offer as the client with the advice of technical advisers is prepared to make, there must be recourse to the arbitration clause of the building contract.

The subject of contractual claims is one that warrants a textbook of its own, and several such have been written. The foregoing remarks therefore are only a basic introduction to the role of the quantity surveyor in dealing with claims.

Bibliography

Aqua Group, The *Contract Administration for the Building Team.* Blackwell Scientific Publications, 1990.

Barrett, F.R. *Cost Value Reconciliation.* Chartered Institute of Building, 1992.

Barrett, F.R. *Financial Reporting – Profit and Provision.* CIOB Technical Information Service No. 12, Chartered Institute of Building, 1982.

Cooke, B. and Jepson, W.B. *Cost and Financial Control in Construction.* Macmillan, 1982.

Fellows, R.F. 'Cash flow and building contractors'. *The Quantity Surveyor,* September 1982.

Hibberd, P.R. 'Variation and NEDO formulae: the non-adjustable element.' *Chartered Quantity Surveyor,* August 1981.

Hibberd, P.R. *Variations in Construction Contracts.* Blackwell Scientific Publications, 1986.

Hudson, K. 'DHSS expenditure forecasting method – Chartered Surveyor'. *B and QS Quarterly,* 1978.

Nicholson-Cole, D. 'Cash flow information for the client'. *Architects Journal,* October 1983.

Nisbet, J. 'Post-contract cost control, a sadly neglected skill'. *Chartered Quantity Surveyor,* January 1979.

Nisbet, J. 'The QS and his work: presentation of final accounts'. *Architects Journal,* October 1972.

Pilcher, R. *Project Cost Control in Construction.* Blackwell Scientific Publications, 1985.

Povall, W.E. 'Post-contract procedures for quantity surveyors'. *Chartered Quantity Surveyor,* July 1981.

RICS *Contractors' Direct Loss and or Expense.* Quantity Surveyors' Practice Pamphlet No. 7, The Royal Institution of Chartered Surveyors, 1987.

RICS *Definition of Prime Cost of Building Works of a Jobbing or Maintenance Character.* RICS Books, 1980.

RICS *Definition of Prime Cost of Daywork Carried Out Under a Building Contract.* RICS Books, 1975.

RICS *Extension of Time under the JCT Standard Form of Building Contract.* Quantity Surveyors' Practice Pamphlet No. 6, RICS Books, 1986.

SCQSLG *Assessment of Liquidated and Ascertained Damages on Building Contracts.* Society of Chartered Quantity Surveyors in Local Government, 1981.

SCQSLG *The Insolvency of Building Contractors.* Society of Chartered Quantity Surveyors in Local Government, 1981.

Somerville, D.H. *Cash Flow and Financial Management Control*. CIOB Surveying Information
    Service No. 4. Chartered Institute of Building, 1981.
Thomas, R. *Construction Contract Claims*. Macmillan Building and Surveying, 1993.
Trickey, G. 'Waging war on contract delays'. *Chartered Quantity Surveyor*, March 1980.
Trickey, G. *The Presentation and Settlement of Contractors' Claims*. Spon, 1983.
Trimble, E.G. and Kerr, D. 'How much profit from contracts goes to the bank?' *Construction News*, March 1984.
Upson, A. *Financial Management for Contractors*. Blackwell Scientific Publications, 1987.
Willis, C.J. and Moseley, L. *More Advanced Quantity Surveying*. Granada, 1983.

Chapter 12

Project management

Introduction

Increasingly, clients are adopting a project culture in all aspects of their business. Project management is not new or specific to construction contracts.

The organization and management of construction projects has existed in practice since buildings were first constructed. The process long ago was much simpler but, as knowledge increased and societies became more complex, so the principles and procedures involved in management evolved. In some countries, notably in the USA, the management of construction works began to emerge as a separate and identifiable professional discipline some years ago alongside those of architecture and engineering. Because of the differences in the way the industry is structured in the UK, the professions have not developed in the same way, nor to the same extent. It is now, however, being accepted by more and more clients that, to succeed in building, someone needs to take the responsibility for the overall management of the construction project. Indeed some clients believe the project culture is slow in being adopted within the construction industry. Project management is a very different function from either design or construction management and requires other, different qualities, which are not necessarily inherent in the more traditional disciplines.

There has in recent years been a considerable interest amongst quantity surveyors in project management of one sort or another. This has been evidenced by the increased number of postgraduate courses, textbooks and other publications on the subject, and of practitioners seeking to specialize in this type of work. Some quantity surveyors, anticipating the possible demise of their traditional role, have seen project management as a source of work for the future. Others suggest that the financial expertise of quantity surveyors makes them ideally suited to such a role.

Definition

Project management in construction has been described as:

> The planning, control and coordination of a project from conception to completion (including commissioning) on behalf of a client. It is concerned with the identification of the client's objectives in terms of utility, function, quality, time and cost, and the establishment of relationships between resources. The integration, monitoring and control of the contributors to the project and their

output, and the evaluation and selection of alternatives in pursuit of the client's satisfaction with the project outcome, are the fundamental aspects of construction project management.' (Walker: *Project Management in Construction*.)

Historical note

The traditional relationships that have existed between the client, consultants and contractors have on occasions become strained. This is due to the increasing complexity of design and construction, the importance of early completion, the need for better building performance, and the increasing concern for financial control in its entirety. It is apparent from these demands that there is an urgent need for a much greater understanding and interpretation of the client's requirements. In addition, there is a necessity for improved communications, together with a closer coordination of the work of all those involved in the design and construction processes. This is in part due to the changing nature of the construction industry and the different procurement routes adopted. Project management provides the necessary link between the client and the consultants, and also offers such a coordinating role.

Title

Whatever name is given to the role of the project manager, whether it be project controller or project coordinator, the idea is that one person should have the overall control and responsibility for coordinating the activities of the various consultants, contractors, subcontractors, processes and procedures for the full duration of the project. The project duration in this context starts at inception and ends on the completion of the defects liability period. The management process may also extend into the time when the building is in use towards whole-life and facilities management.

A project manager should not attempt to perform any of the functions normally undertaken by the design team. These should always be separate, to avoid having to make any compromise decisions that might otherwise occur.

The titles listed above attempt to describe a particular role. They should not, however, be confused with the contractor's project manager, who will primarily manage the construction process.

A further development of the management function is that of the construction manager, whose role involves the management of the design, procurement and construction. No main contractor exists; instead, the construction manager manages the trade contractors, and as reimbursement is by way of a fee this role can be carried out in a more objective manner with due recognition of the client's and the project's interests.

The project manager has not until recently had specific authority under the standard forms of contract, it being the architect/contract administrator's responsibility to administer the contract. GC/Works/1 has now introduced the project manager in a key role with extensive powers under the contract; however, these powers relate more to the administration of the contract and not the overall management of the project.

Requirements for a project manager

Management is essentially a human matter, and this fact overshadows all other considerations. No one professional group therefore has a monopoly of the skills required, although the quantity surveyor is well suited to this role and perhaps better suited than many others. Various individuals have, however, chosen to specialize in management, having emerged from backgrounds in architecture, engineering, building and surveying. Management is also a job, and a highly technical one. The following are the basic requirements for anyone aspiring to the role of project manager.

Personality

There are those who will attempt to argue that good managers are born and not made, and that the inherent qualities must already be present in an individual. They will equally argue that no amount of education and training or even experience can produce good managers. However, improvement in performance can always be achieved by encouragement in the right direction, and hence further study is necessary if the role is to be carried out to the full.

Self-motivation and the ability to motivate others are very important, as is the relationship with those with whom the manager has to work. Own personal goals and moral values, coupled with own attitude to work and that of others, also need to be considered. The response to the various aspects of the project will vary depending upon the project manager's own professional allegiance.

Technical knowledge

The second factor to consider is the technical ability of the person. There is something unique about the construction industry, and project management within it is a rather specialized form of management. The techniques derived from the manufacturing industry, for instance, often do not work on a construction site, as construction is concerned with one-off projects undertaken on the client's premises. Ideally, therefore, the project manager will already be a member of one of the construction professions. An understanding of the process and the product of construction, and a working knowledge of the structure of the industry, will clearly be advantageous, if not essential. The importation of managers with no knowledge of the industry or its workings from outside these traditional disciplines has drawbacks, and their appointment should be approached with caution.

Managerial skills

A third requirement of the project manager is a knowledge of the subject of management: at least a rudimentary knowledge of the principles and practice of management and of the techniques that can be applied. Several management theories exist and the project manager must select and adapt those that are of the most relevance to particular situations.

Qualities of the project manager

It is difficult to be dogmatic about this question because there are no real rules. However, there are certain personal qualities that are both desirable and helpful in executives who have to fulfil the particular processes of management. There is a need, for example, for integrity, and for clarity of expression when speaking and writing. Loyalty, fairness and resourcefulness are also necessary. The following are perhaps some of the more important qualities to be looked for.

Leadership

Whenever a group of people work together in a team, the situation demands that one of the members becomes the leader. The project manager is the designated leader of the construction team, whose duty it is to ensure that the whole work is carried out as efficiently as possible. The responsibility of combining the various human resources and obtaining the best from them is the project manager's, who must seek to complement the attributes of the various members of the team and keep conflict to a minimum.

The quality of leadership is a basic function of project management. The ability to lead and motivate others while commanding their respect is an essential characteristic. A good leader is never ruthless, as this creates conditions of stress, strain and insecurity. The project manager's character and ability will set the tone from the beginning and from this the loyalty of the other members of the team will be gained. The project manager must of course know when to praise and when to reprove, and must also have the courage to admit a mistake, to make changes or to proceed against opposition.

Clear thinking

The ability to think clearly is also an important aspect of management. A confused mind creates confusion around it and confused instructions to other members of the project team are signs that management is incapable of making decisions. The inability in the first place to think clearly originates from an inadequate understanding of the objectives and priorities associated with the problem.

Delegation

The total amount of knowledge required in the management of a project is beyond the scope of a single individual. The manager must therefore be able to delegate certain tasks and duties to others involved in the project, and be able to rely upon and receive advice from them. The inability to delegate results in overwork for the project manager, frustration on the part of others and a generally badly run project. Everyone must feel that they are able to make a valid contribution to the overall management of the project.

Decision-making

Decisions need to be made at all levels in the organization of a construction project. Those made at the top will be concerned more with policy, client objectives and the framework for the project as a whole. At lower levels, they tend to relate more to the solving of particular problems. The aptitude for making decisions is an important quality that distinguishes the manager from the technologist. The general level of complexity of construction projects and the number of consultants involved make the decision-making for the project manager particularly difficult. Making a rapid decision requires a certain amount of courage. Sticking to a decision in the face of criticism, opposition or apparent failure requires a large amount of conviction. It is, of course, vitally important for the project manager to make the right decision, and this can only consistently occur through experience. The ability to sense a situation and exercise correct judgement will always improve with practice.

Duties and responsibilities of the project manager

The project manager's terms of engagement, extent of authority and basis of fee reimbursement must be established prior to appointment. The experienced project manager will realize the importance of unambiguous conditions of appointment.

The duties of a project manager in the construction industry will vary from project to project. Different countries around the world will also expect a different response to the situation, depending on the contractual systems that are in operation. The mistake, and perhaps the reason for the failure, of the traditional system in certain instances is that an attempt is made to use a single system to suit all circumstances. Any contractual arrangement, however good, must be adapted to suit the needs of the client and the project, and not vice versa. The project manager will need to employ a wide variety of skills and options for a whole range of different solutions. The following may therefore be described as some, but not necessarily all, of the duties of the project manager.

Client's objectives

The starting point of the project manager's commission is to establish the client's objectives in detail through discussion. The success of any construction project can be measured by the degree to which it achieves these objectives. The client's need for a building or engineering structure may have arisen for several reasons: to meet the needs of a manufacturing industry, as part of an investment function, or for social or political demands. In an attempt to provide satisfaction for the client, three major areas of concern will need to be considered. The weighting given to these factors will vary depending upon the perception of the client's objectives.

Performance

The performance of the building or structure in use will be of paramount importance to the client. This priority covers the use of space, the correct choice of materials, adequate design and detailing, and the aesthetics of the structure. Attention will also need to be paid to future maintenance requirements once the building is in use. Clients are more likely to be concerned with the functional standards of the project and, to a lesser extent, the aesthetics.

Cost

All clients will have to consider the cost implications of the desired building's performance. The price that they are prepared to pay will temper to some extent the differences between their needs and wants. The importance of construction costs is very often underestimated. Clients today are also more likely to evaluate costs not solely in terms of initial capital expenditure but rather on a basis of life-cycle cost management.

Time

Once clients decide to build they are generally in a hurry for their completed product. Although they spend a large amount of time deliberating over a scheme, once the decision to build has been reached then they often require the project to be completed as quickly as possible. In any event, in order to achieve some measure of satisfaction, and to prevent escalating costs, commissioning must be achieved by the due date.

The project manager's strategy for balancing the above three factors will depend upon an interpretation of the client's objectives. It would appear, however, that there is some room for improvement in all three areas. The improvement of the design's completeness, particularly, should reduce the contract time and hence the constructor's costs. The correct application of project management should be able to realize benefits in these areas.

The client's objectives should be used as a goal for the broader issues involved in the design and construction of the project. The discernment of these objectives will assist the project manager to decide which alternative construction strategies to adopt. It is very important that an adequate amount of time is allowed for a proper evaluation of the client's needs and desires. Failure to identify these properly at the outset will make it difficult for the project to reach a successful conclusion upon completion.

Client's brief

This involves the evaluation of the user requirements in terms of space, design, function, performance, time and cost. The whole scheme is likely to be limited one way or another by cost, and this in turn will be affected by the availability of finance or the profits achieved upon some form of sale at completion. It is necessary therefore for the project manager to be able to offer sound professional advice on a large range of questions, or to be able to secure such information from one of the professional consultants who are likely to be involved with the scheme. This will include the coordination of all necessary legal advice required by the client. It is most important that the client's objectives are properly interpreted as at this stage ideas, however vague, will begin to emerge, and these will often then determine the course of the project in terms of both design and cost.

It is the project manager's responsibility to ensure that the client's brief is clearly transmitted to the various members of the design team, and also that they properly understand the client's aims and aspirations.

Contractor involvement

The client will probably require some initial advice on the methods that are available for involving the contractor in the project. The necessity for such advice will depend upon the familiarity of the client with capital works projects. The correct evaluation of the client's objectives will enable the project manager to recommend a particular method of contractor selection. It may be desirable, for example, to have the contractor involved at the outset or to use some hybrid system of contractor involvement. The project manager will be able to exercise expert judgement in this respect by analysing the potential benefits and disadvantages for the project concerned. This decision will need to be made reasonably quickly, as it can influence the entire design process and the necessity of appointing the various consultants.

Design team selection

The project manager is responsible for the selection and organization of the design team. If the client has been involved in capital works projects previously, they may already have designated consultants with whom the project manager will need to work. The project manager is likely to be responsible for agreeing fees and terms of appointment of all consultants on behalf of the client. In certain instances the project manager may appoint the consultants direct as sub-consultants. Under such circumstances the only contractual link is between the client and the project manager. Where the contractor is to be appointed during the design stage, then the project manager will need to consider the means of selection. Whatever the circumstances, the project manager must control rather than be controlled by either the contractor or any of the consultants. The relationships between and contributions from each consultant must be clear at the outset to avoid any misunderstandings that may occur later.

Feasibility and viability reports

During the early stages of the design process, it will be necessary for the project manager to examine both the feasibility and the viability of the project. Sound professional advice is very important at this stage, as this will determine whether or not the project should proceed. A feasible solution is one that is capable of technical execution and may only be found after some site investigation and discussion with the designers. A feasible solution may, however, prove not to be viable in terms of cost or other financial expenditure. Unless the project is viable in every respect, it will probably not proceed. The investigation work should be sufficiently thorough while taking note of the fees involved, particularly if the project should later be abandoned.

Planning/programming

Once the project has been given the go-ahead it will then become necessary to prepare a programme for the overall project, incorporating both design and construction. The programme should represent a realistic coordinated plan up to the commissioning of the scheme. The project manager must carefully monitor, control and revise it where necessary. Several useful techniques exist for programming purposes, and where these can be computer-assisted rapid updating can easily be achieved. The selection of the appropriate technique will allow the project to be properly controlled in terms of time.

Design process management

Project information is often uncoordinated, and this leads to inefficiency, a breakdown in communications between the design team, frequent misunderstandings and an unhappy client. For example, the delayed involvement of service engineers often results in changes to the design of the structure both to accommodate the engineering work and also to incorporate good engineering ideas. An important task therefore for the project manager is to ensure that the various consultants easily and frequently liaise with each other while maintaining their own individual goals. There is not much room in the design team for those who wish to go it alone; teamwork is very much underrated, but it is vital for the success of the project. The project manager will therefore need to exercise both tact and firmness in ensuring that the client's objectives remain paramount.

The project manager, although not directly involved in the process of designing in its widest sense, must nevertheless have some understanding of design in order to appreciate the problems and complexities of the procedures involved. Responsibility for the integration and control of the work from various consultants rests with the project manager, who must approve the work that they have undertaken, take the full and ultimate responsibility for their work, and be directly answerable to the client for all facets of the project. This will include ensuring quality control of all aspects of the design (and construction) process and carrying out regular technical audits on the developed design solutions.

Clients often feel that they are not properly informed about the work of the consultants, and the responsibility for ensuring that they are lies with the project manager.

The project manager must also be kept informed of the cost implications as the design develops. The design team must be advised of what can or cannot be spent. In this respect the control of the costs should be more effective than when relying upon the efforts of the architect alone. The project manager must, of course, have a very clear understanding of the client's intentions in this and will also need to advise the client in those circumstances where the original requirements cannot be met in terms of design, cost or time. The project manager will always have an eye on the future state of the project and must keep at least one step ahead of the design team.

During this stage, unless the contractor has been appointed earlier, the project manager will need to consider a possible list of firms who are capable of carrying out the work.

Throughout this time the project manager will also need to ensure that the proper and timely action is taken to obtain all statutory approvals.

Supervision and control during construction

The project manager should try to make sure that the design of the works is as near complete as possible prior to tendering. This is likely to result in fewer problems on site, a shorter contract period with a consequent reduction in costs, and commissioning at the earliest possible date. During the contract period the project manager will need to have regular meetings with the consultants and the contractor and his subcontractors. Progress of the works must be monitored and controlled and any potential delays identified. The effect on the programme and the budget for any variations will also need to be monitored. The project manager must be satisfied that the project is finished to the client's original requirements; although one of the consultants may be responsible for the quality control, the project manager will need to be careful about accepting substandard or unfinished work. Some problems may need to be discussed with the client, but early decisions should be sought to bring the project to a successful conclusion. The project manager might have an ongoing role after the main construction contract to administer fitting-out work for occupiers and tenants.

Evaluation and feedback

This represents the final stage of the project manager's duties. It should be ascertained that all commissioning checks have been carried out satisfactorily, that the accounts have been properly agreed and that the necessary drawings and manuals have been supplied to the client. The project manager will need to advise on current legislation affecting the running of the project, on grants, taxation changes and allowances. It may also be necessary to 'arbitrate' between consultants and contractors in order to safeguard the client's interests. The client should be issued with a 'close-out' report to identify that responsibilities of all parties have been satisfactorily discharged, and this will also assist in any future capital works project that the client might undertake.

Quantity surveying skills and expertise

The skills of the quantity surveyor traditionally included measurement and valuation and to these were later added accounting and negotiation. As the profession evolved, these skills were extended to include forecasting, analysing, planning, controlling and evaluating. The modern-day quantity surveyor has increased the expertise even further to include communications, budgeting, problem-solving and modelling. Knowledge has also been considerably developed both by a better understanding of the design and construction process and by having a broader base.

This provides the quantity surveyor with an excellent background, which is appropriate for project management. Indeed, a significant number of those already engaged in this work are members of the quantity surveying profession. The traditional role of the consultant quantity surveyor is often seen as advisory rather than managerial. In essence, therefore, the difference between the traditional role of the quantity surveyor and that of the project manager lies in his attitude and method of approach.

Fees

In most countries, including the UK, the fees charged for professional services include the management of projects for clients.

If the service being offered is enhanced by project management, then some extra charge will be deemed equitable. As the existing poor management organization is to some extent responsible for the poor quality, time delays and extra costs, then a process that attempts to rectify these problems should be worth paying for. The professional fee involved may in any case show a saving to the client overall. The type of management envisaged was never deemed to be included in the existing traditional fee structures.

The project manager will in the first instance need to negotiate a fee with the client. This fee will need to take into account the type of service being offered, the complexity of the project, its volume and duration. In periods of heavy competition, clients tend to require lump sum fees to be agreed for project management services. It is essential therefore that the basis of the fee in terms of scope and time-scale is clearly stated in the terms of appointment.

Education for project management

Quantity surveying education is constantly evolving to meet the needs of the profession and industry of the future. Subjects that were once important have now been relegated to a position of low importance, as new subjects vie for space in the curriculum.

Quantity surveying undergraduate courses now include a study of project management, with the view that many graduates may be practising this in the future. For those surveyors who have not had any formal education in management, attendance at short courses and seminars should prove worth while. In other circumstances it may be desirable to study for a diploma in management studies or, if time allows, a master's course in construction or project management. In order to be effective in this area some formal study is desirable.

Bibliography

Bennett, J. *Construction Project Management*. Butterworth-Heinemann, 1985.
Bennett, J. *International Construction Project Management*. Butterworth-Heinemann, 1991.
Briner, J., Geddes, B. and Hastings, R. *Project Leadership*. Gower, 1990.
CIOB *Code of Practice for Project Management in Construction and Development*. Chartered Institute of Building, 1992.
CIOB *Project Management in Building*. Chartered Institute of Building, 1982.
Davis, N. 'The consultant QS and project management'. *Chartered Quantity Surveyor*, November 1983.

Fisk, E.R. *Construction Project Management.* Prentice-Hall, 1992.

Fryer, B. *Construction Management: Principles and Practice.* Blackwell Scientific Publications, 1990.

Harris, F. and McCaffer, R. *Modern Construction Management.* Blackwell Scientific Publications, 1989.

Haynes, M. *Project Management: From Idea to Implementation.* Kogan Page, 1990.

Hibberd, P., Merryfield, H. and Taylor, G. *Key Factors in Contractual Relationships.* The Royal Institution of Chartered Surveyors, 1991.

Lock, D. *Project Management Handbook.* Gower, 1987.

Lock, D. *Project Management.* Gower, 1988.

Pilcher, R. *Project Cost Control in Construction.* Blackwell Scientific Publications, 1985.

Reiss, G. *Project Management Demystified: Today's Tools and Techniques.* Spon, 1991.

RICS *Construction Management and the Chartered Quantity Surveyor.* The Royal Institution of Chartered Surveyors, 1986.

RICS *Project Management – Conditions of Engagement,* 2nd edn. RICS Books, 1992.

RICS *Project Management Agreement and Conditions of Engagement: Guidance Notes,* 2nd edn. RICS Books, 1992.

Walker, A. *Project Management in Construction.* Blackwell Scientific Publications, 1989.

Chapter 13

Litigation and dispute resolution

Introduction

From time to time quantity surveyors find themselves involved in litigation either in the courts, in arbitration or in alternative dispute resolution cases (ADR). Their involvement is often as witnesses of fact: that is, someone who was actually there at the time as project surveyor or manager. More often, however, they are involved as expert witnesses, adjudicators, arbitrators themselves or as neutrals or mediators in ADR cases. Each of these roles is considered in this chapter under the following headings:

- arbitration
- alternative dispute resolution
- adjudication
- expert witness work
- lay advocacy.

Arbitration

Disputes between parties to a contract are traditionally heard in the courts, but in building contracts the chosen method is more often arbitration. Arbitration to resolve disputes in building contracts comes about following the agreement of the parties, either when the dispute arises, or more often as a term of the original contract. For instance, the JCT forms of contract all provide that if a dispute arises between the parties to the contract then either party can call for arbitration. When such clauses exist in contracts the courts, if asked, will generally rule that arbitration, having been the chosen path of the parties, is the proper form for the dispute to be heard. Recent legislation (Courts and Legal Services Act 1990) does provide that the courts may hear such disputes if both parties, having changed their minds, are in agreement, but such cases are likely to be rare.

The JCT publish arbitration rules for use with arbitration agreements referred to in the JCT contracts. They set out rules concerning interlocutory (intermediate) matters, conduct of arbitrations and various types of procedures: without hearing (documents only), full procedure with hearing, and short procedure with hearing, each containing strict timetables.

Arbitration and the law relating to it are subjects on their own, and the bibliography at the end of this chapter includes selected reading on the subject. The traditional advantages of arbitration over the courts were fourfold:

- Arbitration proceedings are quicker than the courts.
- Arbitration is cheaper than litigating in the courts.
- The parties get a judge of their choosing, a person knowledgeable about the subject matter in dispute but with no knowledge of the actual case, rather than a judge imposed upon them.
- Unlike proceedings in the courts, arbitration proceedings are confidential.

With regard to the first of these traditional advantages, provided the proceedings are kept simple then arbitration can still prove quicker than proceeding through the courts. However, the modern tendency to involve lawyers at all stages and to complicate disputes has eroded this advantage, and there is now often very little difference between the time it takes to get a dispute settled either in arbitration or in the courts. Equally, arbitration is no longer the cheaper option that it used to be. Legal costs, and the fact that the parties have to pay for the judge and courtroom in arbitration proceedings instead of having them provided at the taxpayer's expense, have eroded this traditional advantage as well.

However, the third and fourth advantages – choice of judge and confidentiality – still exist and are considered by many to be of overriding importance: hence the survival and indeed increase in the number of arbitration references.

The duty of arbitrators is to ascertain the substance of the dispute, to give directions as to proceeding, and as quickly as possible to hear the parties and make their decision known by way of an 'award'. Arbitrators hear both sides of an argument, decide which they prefer and award accordingly. They cannot decide that they do not like either argument and substitute their own solution; this would lead to an accusation of misconduct (see below).

An arbitrator may be named in a contract, although this is rare; it is usually thought better to wait until the dispute arises and then choose an appropriate person. The choosing will be by the parties; each side exchanges suitable names and usually an acceptable choice emerges. If the parties are unable to agree then a Presidential appointment will be sought from a body such as the Chartered Institute of Arbitrators, the Royal Institute of British Architects or The Royal Institution of Chartered Surveyors. If a quantity surveyor is thought the most appropriate choice then the President of the RICS would be the most likely to be approached.

When appointed, an arbitrator calls the parties together in a preliminary meeting. At this meeting the nature and extent of the dispute are made known to the arbitrator and an order is sought for directions fixing a timetable for submission of the 'pleadings' (points of claim, points of defence, etc.) The directions normally end by fixing a date and venue for a hearing.

As the date for the hearing approaches, the parties will keep the arbitrator informed of the progress they are making in working their way through the timetable. If a compromise is achieved and the matter is settled, they will inform the arbitrator immediately.

At the hearing each party puts their case and then calls their witnesses: first the witnesses of fact and then the expert witnesses. All witnesses normally give evidence under oath and are examined and cross-examined by the parties' advocates.

All arbitrators are bound by the terms of the Arbitration Acts 1950 and 1979. Their awards, once made, are final and binding. If an award is not honoured then the aggrieved party can call on the courts to implement the award. The only exception to this is if the arbitrator is found to be guilty of misconduct, or the arbitrator is wrong at law.

If either of these events occurs a party can apply to the courts either to have the arbitrator's award referred back for reconsideration or, in extreme cases, to seek the removal of the arbitrator.

Being wrong at law is self-explanatory. The arbitrator may be a lawyer but more likely does not hold that qualification; there is therefore no disgrace in getting the law wrong, although if in doubt an arbitrator has the power under the Act to seek legal assistance and is often well advised to do so.

Misconduct is more difficult to describe. It is not misconduct in the usual understanding of the word; it can be defined in this context as a failure to conduct the reference in the manner expressly or impliedly prescribed by the submission or to behave in a way that would be regarded by the court as contrary to public policy. Failure to answer all the questions asked, hearing one party without the other party being present, making mathematical errors: all these are examples of misconduct that the courts would consider and direct as they thought fit.

Quantity surveyors are well suited to act as arbitrators in building disputes as the matter invariably involves measurement, costs and loss and/or expense and interpretation of documents: all matters falling within the expertise of a quantity surveyor. When the matters in dispute concern quality of workmanship or design faults then arbitration is best left to architects or engineers. Equally, when matters of law are the prime consideration then the arbitrator should be a lawyer or at least have a law qualification.

Alternative dispute resolution

Alternative dispute resolution, or ADR as it is generally known, originated in the USA, was adopted in Britain in the 1980s and is now practised worldwide.

ADR provides a means of resolving disputes without resorting to arbitration or the courts. In that respect it is nothing new; quantity surveyors and contractors have over the years traditionally settled disputes by negotiation. With ADR the process is somewhat more formal.

The advantages of ADR can be summarized as follows:

- *private*: confidentiality is retained;
- *quick*: a matter of days rather than weeks, months or even years;
- *economic*: legal and other costs resulting from lengthy litigation are avoided.

None of these advantages will be achieved, however, unless one vital ingredient is present. There must be goodwill on both sides to settle the matter on a commercial rather than a litigious basis. If this goodwill does not exist then the parties have no option but to resort to arbitration or the courts, without wasting further time and resources.

ADR can take a variety of forms

- mini-trial
- mediation
- mutual fact-finding
- mutual expert
- private judging.

Of these, the first two are the most commonly met.

Mini-trial

Each of the parties is represented, generally but not necessarily by a lawyer, who makes a shortened presentation of their client's case to a tribunal. This presentation will have been preceded by limited disclosure of documentation, or 'discovery' as it is known. The purpose of this limited discovery is to ensure that each party is aware of the opposite side's case and is not taken by surprise by the presentation.

The tribunal usually takes the form of a senior managerial representative of each party and an independent advisor known as a 'neutral'. It is important to the success of the mini-trial that each of the parties' representatives should not have been directly involved in the project, and is therefore able to be less emotionally involved than someone who has lived with the dispute and has difficulty in taking a detached view. It is also essential that both parties' representatives have full authority to settle the matter. There is nothing worse than arriving at what appears to be consensus for one party then to disclose that they can only agree subject to authorization from a chairman, board, council, chief officer or some other third party.

The neutral has to be someone with a knowledge of the industry but no knowledge or interest in the dispute. In this respect the same requirements as that of a good arbitrator are called for. However, unlike an arbitrator, whose task it is to hear the arguments and decide which is preferable, a neutral becomes much more involved, listening, suggesting, and giving advice on matters of fact and sometimes on law as well.

After the initial presentation, experts and witnesses of fact may be called, following which the managers enter into negotiation with a view to coming to a consensus. The length of these negotiations will depend very much on the complexity of the matters in dispute and they will be assisted by the neutral, who may if the parties remain deadlocked give a non-binding opinion, which may lead to a settlement after further negotiation.

Once a negotiated settlement is reached, the neutral will there and then draft the heads of a statement of agreement which the parties will each initial. This will then be followed by a formal agreement ending the matter.

Mediation

This is a less formal method of proceeding than that described for a mini-trial. The parties, with assistance from their experts or lawyers, will select a neutral whose background will reflect the matters in dispute. A preliminary meeting will be called by the neutral to find out the substance of the dispute and to decide how best to proceed.

At a subsequent meeting the parties make formal presentation in joint session. There then follows a series of private meetings, or 'caucuses' as they are termed, between the neutral and each of the parties on their own. The neutral moves from one caucus to the next reporting (with agreement) the views of each party in turn. This should lead to the neutral's being able to suggest a formula for agreement, which in turn may lead to a settlement. Such agreement is terminated in the same way as in a mini-trial.

Mutual fact-finding

Resolution of a dispute by mutual fact-finding is an informal procedure whereby the parties, possibly at a different level from those closely involved, take a pragmatic and commercial approach to settling the dispute with or without the assistance of a mutual expert.

Conclusion

To summarize, ADR provides a dispute resolution mechanism that concentrates on resolving disputes by consensual rather than adjudicative methods. To quote from the Centre for Dispute Resolution's introduction to ADR:

> These techniques are not 'soft' options but rather involve a change of emphasis and a different challenge: how to get the best possible settlement rather than trial result. The parties cooperate in the formulation of a procedure and result over which they have control. The main features are easily captioned under the '4C's'

> - Consensus – a joint objective to find the business solution;
> - Continuity – a desire to find a solution in the context of an ongoing business relationship;
> - Control – the ability to tailor a solution that is geared towards a business result rather than a result governed by the rule of law, which may be too restrictive or largely inappropriate;
> - Confidentiality – avoiding harmful washing of dirty linen in public.

Quantity surveyors have a role to play in ADR. They are obvious candidates for the post of neutral in building disputes, provided they have the right negotiating skills and have undergone the necessary training.

Adjudication

Adjudication is the giving of a non-binding opinion at the request of a party or parties to a dispute.

Adjudication clauses will be found in certain contracts. For instance the JCT Nominated Sub-Contract Conditions (NSC/C(PCC)) contain clauses providing for an adjudicator (who was named in the contract conditions when the contract was signed) to give a ruling regarding the amount of 'set-off' (deduction made by a main contractor from a sum certified to be paid to a nominated sub-contractor). In this instance the adjudicator has the power to order either:

- that a set-off sum should stand;
- that a certain sum should be paid; or
- that the disputed amount be paid into a stakeholder's account.

The ruling then stands until such time as the parties settle their differences or an arbitrator makes an award.

Adjudication clauses will also be found in the ACA Form of Building Agreement and in the Government GC/Works/1 contract. In this latter case the adjudicator is an officer of the authority and has power to consider any matter in dispute other than those described in the contract as final and binding decisions of the project manager. In this form the adjudicator can only act after a six-month 'cooling off' period, which is surprising in that adjudication is intended to provide a quick solution to a dispute and not to let the matter fester on. Again, the adjudicator's decision is binding until overtaken by an arbitrator's award.

One of the benefits of adjudication is that it can often lead to a settlement without the matter going any further. This is because a party that has lost in an adjudication will think very carefully before proceeding with very expensive litigation or arbitration where they might well lose again with the additional penalty of paying the other side's costs as well.

Expert witness

Quantity surveyors, particularly those more experienced, may be briefed to give expert witness in a variety of circumstances:

- litigation in the High Court or in the County Court
- in arbitration
- before a tribunal.

While proceedings in court tend to be more formal than in arbitration, the same rules apply.

Expert evidence is evidence of opinion put forward in support of a particular cause by an expert, who must be seen at all times to be independent and who has no financial interest in the outcome. Expert witnesses are there to assist the court, the arbitrator or the tribunal and must not be seen to be solely advocating their client's case. To emphasize this, two judgmental extracts are relevant.

In the case of *Whitehouse* v. *Jordan* (1981) 1 WLR 247 Lord Wilberforce warned:

Whilst some degree of consultation between experts and legal advisers is entirely proper it is necessary that expert evidence presented to the court should be and should be seen to be the independent product of the expert and uninfluenced as to form or content by the exigencies of litigation. To the extent that it is not the evidence is likely to be not only incorrect but self-defeating.

The second quotation comes from the judgment of Sir Patrick Garland in the case of *Warwick University* v. *Sir Robert McAlpine and Others* (1988) 42 BLR:

> It appeared to me that some (but by no means all) of the experts in this case tended to enter into the arena in order to advocate their client's case. This led to perfectly proper cross-examination on the basis: 'You have assembled and advanced explanations which you consider most likely to assist your client's case'. It is to be much regretted that this had to be so. In their closing speeches counsel felt it necessary to challenge not only the reliability but the credibility of experts with unadorned attacks on their veracity. This simply should not happen where the court is called upon to decide complex scientific and technical matters.

An expert opinion must be based on facts, and these facts have to be proved unless they can be agreed. The process of agreement is achieved by meetings of the experts, frequently ordered by a judge or arbitrator to agree facts and figures wherever possible, and to narrow the issues in dispute. These meetings are usually described as being 'without prejudice': that is, nothing discussed can be used in evidence. Such meetings are followed by a joint statement of what has been agreed and of those matters that are still outstanding. In fact, such meetings can often lead to a settlement of the dispute without troubling the tribunal further.

Acting as an expert witness is a time-consuming and therefore an expensive exercise. There needs to be careful reading of all relevant matters, some research where appropriate and a studiously prepared proof of evidence. As has been said above, such proofs of evidence have to be seen as being in the sole authorship of the expert; while they may listen to advice, the final document (on which they stand to be cross-examined) is theirs and theirs alone.

Lay advocacy

Advocacy in the High Court and in the County Courts is restricted to barristers and solicitors. There is no such requirement when it comes to arbitration or appearing before a lay tribunal; anyone can advocate their own or their client's case.

It sometimes happens that when a matter is strictly technical, such as a dispute on measurement or computation of a final account, an arbitrator may direct that the two quantity surveyors should appear before him or her to present their client's case. When this happens, quantity surveyors are acting as lay advocates and are dealing with matters very much within their own expertise. Occasionally it is suggested that a quantity surveyor might act as lay advocate on the whole case, not just that part of it within his or her own discipline.

Advocacy is a skilled art, that is not easily learned. It involves meticulous preparation, the ability to ask the right questions in the right way, and the ability to think on one's feet and respond quickly to answers from witnesses or points made by one's opponent. For this reason quantity surveyors should consider very carefully whether or not to accept an invitation to act as advocate, and unless they are quite confident that they are able to, should politely decline.

Bibliography

Arbitration

Bernstein, R. *Handbook of Arbitration Practice* 2nd edn. Sweet & Maxwell, 1993.
Bird, R. *Osborn's Concise Law Dictionary* 8th edn. Sweet & Maxwell, 1993.
Fay, E. *Official Referee's Business* 2nd edn. Sweet & Maxwell, 1988.
Hawker, G., Uff, J. and Timms, C. *ICE Arbitration Practice*. Thomas Telford, 1985.
Jones, N.F. *User's Guide to the JCT Arbitration Rules*. Blackwell Scientific Publications, 1989.
Marshall, E.A. *Gill's Law of Arbitration* 3rd edn. Sweet & Maxwell, 1981.
Mustill, M.J., and Boyd, S.C. *Commercial Arbitration*, 2nd edn. Sweet & Maxwell, 1989.
Parris, J. *Arbitration: Principles and Practice*. Blackwell Scientific Publications, 1985.
Powell-Smith, V. and Sims, J.H.M. *Construction Arbitrations*. Legal Studies and Services, 1989.
RIBA *Architect as Arbitrator*. RIBA Publications, 1987.
Stephenson, D.A. *Arbitration Practice in Construction Contracts*. E. & F.N. Spon, 1987.
Walton, A. *Russell on The Law of Arbitration* 22nd edn. Stevens, 1994.

Alternative dispute resolution

Acland, A. *Sudden Outbreak of Common Sense*. Hutchinson Business Books.
The ADR Route Map. Centre for Dispute Resolution.
Brown H. and Marriott A. *ADR Principles and Practice*. Sweet & Maxwell, 1993.

Expert witness

Cross, R. and Tapper, C. *Cross on Evidence*. Butterworth, 1990.
Mildred, R.H. *The Expert Witness*. Longman, 1982.
Reynolds, M.P. and King, P.S.D. *The Expert Witness and His Evidence*. Blackwell Scientific Publications, 1992.

Chapter 14

Special client services

Introduction

The ever-changing nature of the construction industry, the influence of alternative procurement routes, and the developments in information technology, have all had an impact on the role of the quantity surveyor. This has led to a wide diversification of the services provided by quantity surveying practices in addition to the more traditional aspects of cost management, procurement advice, document production and contract administration. This has sometimes resulted in specialist staff being employed to carry out specific functions.

Many of these specialist services have developed out of, and benefit from quantity surveyors' specific skills, such as a detailed understanding of contractual issues, negotiation skills and a global understanding of construction-related issues. It is considered by some that it is these specific skills that have enabled quantity surveyors to apply themselves to new areas of business activity with considerable effect. The diversification of specialist services offered is likely to increase as practices identify new markets and new applications for existing skills that have been developed through education and training.

Two major areas of specialization that quantity surveyors have very much become involved with over the years are project management and litigation and dispute resolution services, both of which warrant consideration in detail and have been covered in Chapters 12 and 13.

Other specialist services that have developed and are now provided by quantity surveyors to a varying extent include:

- grants, capital allowances and tax advice
- insolvency services
- facilities management
- information technology consultancy
- quality assurance consultancy
- development appraisals
- funding advice
- technical auditing
- valuation for fire insurance
- fire loss assessment
- schedules of condition
- dilapidations.

Grants, capital allowances and tax advice

The quantity surveyor is well placed to advise clients on the availability of development grants and other incentives as well as the taxation implications of construction. In many instances accountants alone do not have the necessary detailed knowledge of the construction industry in general nor of any project in particular to advise their clients of the intricacies and application of tax in this context. Proper accounting for grants and tax incentives on a scheme through a correct interpretation of the appropriate legislation can turn a potentially non-viable project into a financially successful one.

Similarly, it is important to have a clear understanding of which specific items of capital expenditure on a construction project qualify for capital allowances that can be claimed against corporation tax. Such allowances apply either to whole buildings such as scientific research buildings and hotels, or to specific items such as 'plant and machinery'. This understanding will initially enable advice to be given to clients during the design process to ensure that qualifying items are incorporated from the outset and also to enable capital allowance claims to be prepared and presented to a company's accountants or directly to the Inland Revenue.

It is important with all advice of this kind that the quantity surveyor is aware of the ever-changing legislation, regulations and current case law when advising clients on these matters.

Insolvency services

If a contractor or property developer becomes insolvent it is important that action is taken quickly. In such circumstances an intimate knowledge of construction costs can be crucial in establishing the true financial picture. The quantity surveyor is therefore well placed to provide the necessary advice, and at a time when there are a large number of insolvencies the number of practices offering these services has increased.

The advice given as part of such a service is again broadly based around the specific skills of the quantity surveyor and might also include project management or construction management, depending upon the circumstances and extent of a particular commission. Insolvency advice services to insolvency practitioners and financial advisers can take a variety of forms.

- *Project appraisals and overviews*: where construction costs and development appraisals are commented upon and confirmed, or work in progress is verified.

- *Corporate rescues*: where a company is in administration with a likelihood of recovery, construction contracts are reviewed, completion budgets and expenditure forecasts are carried out and cost-control systems are established. Post-rescue performance monitoring might also be carried out.

- *Receiverships, administration and liquidation*: where the saving of a company is unlikely, current contracts are reviewed, advice is given on whether schemes should be completed, abandoned, or assigned, and possibly also outstanding claims are negotiated and settled. Project management or construction management services might also be provided.

Facilities management

Facilities management has been defined as 'the practice of coordinating the people and the work of an organization into the physical workplace'.

The components that make up facilities management are not new; however, it is the bringing together of all these components that best describes facilities management. It is a comprehensive service, and includes the management of building, accommodation, premises and estate. It can be divided into four broad areas:

- long-term asset management and planning of capital expenditure;
- building operation management: energy management, space management, maintenance, communications, information technology;
- managing capital projects and staff movements;
- providing support services such as catering, parking and security.

The precise scope of any specific facilities management commission is obviously dependent upon the individual client's requirements.

It can be seen that the facilities management function involves all the principles of management, including planning, organizing, staffing, directing, controlling and monitoring.

The cost of property, support services, IT and the like can amount to a considerable proportion of the overall cost of operating a business and as a result is of significance to senior management. Through careful control of the budgets by the facilities manager considerable savings can often be achieved, and the role is therefore an important one.

A number of surveying-based practices (including quantity surveying) are now offering a facilities management service to clients, who are looking to contract out their non-core business management activities. It is important, however, if a firm is embarking on the provision of a facilities management service, that it has staff with the necessary wide-ranging management skills and expertise, and that it fully understands the client's business and specific requirements in all aspects of the facilities management function.

IT consultancy

Quantity surveying practices have for a number of years been developing IT programmes for their own specific uses, as well as making them available on a commercial basis, which enables the development costs to be recovered and possibly an area of profitable business to be created. Examples of such programmes include computerized measurement and billing, cost planning and library systems.

Certain practices have developed this aspect of their business base to the extent that they combine IT and quantity surveying skills to offer an IT consultancy service to other professionals and clients within the construction industry, and develop software and IT systems on their behalf.

QA consultancy

The development of quality management systems by quantity surveying practices for their own use has in certain instances opened up an opportunity of providing a quality assurance consultancy service to other professionals and client bodies within the construction industry. Firms registered by the certifying bodies such as BSI have developed tailored systems to meet other firms' specific individual requirements, based on the experience of developing and operating their own system.

Development appraisals

The preparation of developer's budgets and other matters concerned with development appraisals for proposed projects are services that are carried out by some quantity surveyors. Clients are themselves becoming more critical and demanding of the process that is used and expect a more wide-ranging type of advice, particularly at the outset or at the more strategic planning level. Other clients already require financial projections, cash-flow analysis and measures to test the sensitivity and reliability of such advice. Appraisals must also take into account the client's operational costs and other matters that are relevant to economic analysis.

Funding advice

Quantity surveyors are able to advise their clients on the sources and methods that can be used for financing the project. It may be necessary to work closely with other financial consultants and fund managers but it is important that the quantity surveyor retains control over the advice that is given. Project funding forms part of the overall package of development appraisal.

Technical auditing

Quantity surveyors' intricate knowledge of contractual matters and the financial aspect of construction mean they are suitable for this role.

It is often assumed that the main function of an audit, whether it is of a company balance sheet, a profit and loss account, or the final account for a construction project, is to detect errors or fraud. This is an incorrect assumption; such detection forms only a subsidiary objective within the main function.

The audit of a final account, or indeed any other account, involves the examination of the account together with any supporting documentation to enable the auditor to report that the account has been properly and fairly prepared so as to give a true and fair view of the state of the contract, according to the information available.

In order to be satisfied in this context the technical auditor must not only compare the contract documentation with the final account but must also examine available records, discuss aspects with the staff concerned, probably visit the site and examine the procedures which have been used. This will involve a more or less complete examination of all the transactions of the contract and the manner in which the accounting work was recorded. The auditor will not, however, try to re-prepare the final account. The extent of the examination of the documents will depend upon the auditor's assessment of internal control and the procedures that should have been followed.

As described in Chapter 5, the formal auditing of a company's accounts is carried out by qualified accountants; however, the needs of a technical audit require someone with a more specialized knowledge of construction procedures, and as it is a financial operation the quantity surveyor is particularly fitted to carry out this function.

Valuation for fire insurance

As a building should be insured for the cost of re-erection in the event of a total loss, the same principles of approximate estimating as set out in Chapter 8 will apply if the quantity surveyor is asked to advise on insurance value. An estimate will be prepared in exactly the same way as it would be for a new building, allowance being made for removal of debris and credit given for salvage. The valuation must include the professional fees involved in rebuilding, and any other charges that the owner or insurance company may have to pay.

A large proportion of fires do not involve a total loss, so it might be thought unnecessary to insure against entire rebuilding. However, even foundations, which it might be thought could be reused, may well be damaged, and for buildings of some age the foundations would almost certainly be insufficient to comply with modern standards. The decision to take any risk must, of course, be with the client, but it should be pointed out that a total loss may be incurred, and that if the full value is not insured then the client must be prepared to be at risk for the balance.

If there are no drawings the surveyor will have to take measurements to enable the cost of rebuilding to be estimated. The usual requirement of insurance companies is that separate values must be put on buildings in different uses, as the rate of premium depends on the use to which the buildings are put. Buildings not of normal brick, concrete or stone construction with slate or tile-covered roofs must also be separately valued, as their premium too may be assessed at a different rate.

It is, of course, important to distinguish valuation for fire insurance, which depends on cost of rebuilding, from valuation for purchase or mortgage, which depends on market value; this latter is outside the normal province of the quantity surveyor.

Fire loss assessment

In assessing fire damage to buildings the surveyor may be acting for the insurance company or for the claimant, and in either case will make a point of visiting the scene of the fire as soon as possible.

If the surveyor is acting for the insurance company, notes of the condition of the premises will have to be taken and sufficient information collected to arrive at an approximate estimate of the value of damage, which the company are fairly certain to require immediately in anticipation of the report. The report to them should include an opinion as to the cause of the fire, which will be formed as a result of questioning anybody available and after examining the debris. An extract from the policy will have been provided and it will be necessary to confirm that the premises damaged are those referred to in the policy, and to see that their use was the same. If the building should have been separately insured, because of either its special construction or its use, this should be stated in the report. Comment on any serious undervaluation in the sum insured will also be made.

If a surveyor is appointed to act for the insured, a meeting will be arranged to negotiate a settlement in much the same way as for a schedule of dilapidations, and a recommendation will be made to the client accordingly. It is not good policy for an insurance company to try to cut a claim to the bone; it is from their reputation for fair dealing in claims that they get their business.

When acting for the claimant the surveyor is, of course, out to get as favourable a settlement as possible for the client. However, the basis of all insurance is good faith, and the company is probably quite ready to meet honestly an honest claim and to give the benefit of the doubt, when there really is doubt.

The assessment of fire damage naturally has a specialized side, as some knowledge of insurance business is necessary. It also, however, requires the special knowledge of the quantity surveyor in measuring and valuing building work, and in important cases may well involve the cooperation of the agent with the quantity surveyor.

Schedules of condition

For new leases of substantial properties it is quite common to prepare a 'schedule of condition' of the property, setting out the condition of the premises at the beginning of the lease. This facilitates defence to a claim for dilapidations at the end of the lease, which would otherwise be based on recollected or imaginary conditions. It is naturally the lessee who initiates steps for its preparation and, if possible, it should be agreed by the landlord. If formal agreement is not obtained, a copy should be delivered to the landlord. The preparation of such a schedule may be regarded as the work of a building surveyor, as no measurement or valuation is involved. It is, however, obviously an advantage for the same person to prepare the schedule and settle the dilapidations.

The schedule of condition should be arranged room by room, setting out the type of finishings and decoration, and noting particularly any defects that might later be alleged to be the lessee's responsibility.

Dilapidations

The usual provision of a repairing lease, that a lessee shall keep the structure in repair and redecorate at definite intervals, is responsible for a good deal of argument and negotiation. Lessees at the end of their lease do not want to be bothered with carrying out repairs and decorating, with the possibility of the lessor having further complaints after the work is done, so are prepared to make a cash payment to settle their liability.

The subject is usually opened by a notice from the landlord requiring that certain repairs set out in a schedule should be done, or payment of damages made for breach of covenant. The schedule will have been prepared by a surveyor, acting for the landlord, who under the terms of the lease has made an inspection of the property. The lessee will appoint a surveyor, who will examine the schedule of the property, make a valuation, and meet the lessor's surveyor to negotiate a settlement. In most cases agreement is reached and reported, and a cash payment is made accordingly. Instructions to deal with schedules of dilapidations usually come from solicitors, who serve the formal notice required by the lease, or whose advice is sought by a lessee receiving a formal notice.

The subject needs some special study of the relevant law, and its detail is outside the scope of this book, but as measurement and pricing are a major part of the work involved, this service may be regarded as one for which the quantity surveyor is qualified.

Bibliography

Alexander, K. *Facilities Management: The Professional's Guide to the State of the Art in International Facilities Management.* Haigh Hilton, 1993.
Ashford, J. *Management of Quality in Construction.* Spon, 1989.
Ashworth, A. 'The auditing of building contracts'. *QS Weekly,* 14th December 1979.
Baum, A. and Crosby, N. *Property Investment Appraisal.* Routledge, 1994.
Birnie, J. 'Energy conservation in buildings'. *Chartered Quantity Surveyor,* April 1983.
Davies, C. 'Major consultants: the quantity surveyor'. *Architects Journal,* April 1984.
Davis, Langdon and Everest *QS 2000: The Future of the Chartered Quantity Surveyor.* The Royal Institution of Chartered Surveyors, 1991.
Fox, R. *Making Quality Happen: Six Steps to Quality Management.* McGraw-Hill, 1992.
Gilbert, G. and Richardson, P. 'Investment appraisal'. *Chartered Quantity Surveyor,* August 1984.
Griffiths, A. *Quality Assurance in Building.* Macmillan, 1990.
Kelly, J. and Male, S. *The Practice of Value Management: Enhancing Value or Cutting Costs.* Gower, 1991.
Livingstone, W. and Easton, C. 'Audit and the quantity surveyor'. *Chartered Quantity Surveyor,* May 1983.
Lumby, S. *Investment Appraisal and Financial Decisions.* Van Nostrand Reinhold, 1991.
Mills, C.A. *Quality Audit: A Management Evaluation Tool.* McGraw-Hill, 1989.
Mohar, J. *Facilities Management Handbook.* Van Nostrand Reinhold, 1990.
RICS *Quality Assurance: Guidelines for the Interpretation of BS 5750 for Use by Quantity Surveying Practices and Certification Bodies.* The Royal Institution of Chartered Surveyors, 1990.
Taylor, M. and Hosker, H. *Quality Assurance for Building.* Longman, 1992.
Upson, A. *Financial Management for Contractors.* Blackwell Scientific Publications, 1987.
Yates, A. and Gilbert, G. *The Appraisal of Capital Investment in Property.* RICS Books, 1989.

Chapter 15

Research and development

Introduction

The wider role of quantity surveying is concerned with the best use of resources. Traditionally much of its practices have been applied to the construction industry, although the techniques and skills have in some cases received a wider application. Many of these practices and procedures have been developed from a pragmatic approach to meet the needs and requirements of clients and the construction and property industries. While there is much that can be described as research and development (R & D), little has been carried out in a structured manner and then only by relatively few members of the profession. However, R & D work is now seen by some of the larger professional practices to be an important part of their activities and, in certain cases, is now a distinctive feature of a practice's profile and portfolio of work. Work within the profession:

- provides a framework within which quantity surveying R & D is encouraged to take place;
- raises an awareness amongst quantity surveyors of the importance and role of R & D;
- further develops a dynamic R & D community;
- seeks to persuade government and other agencies of the importance of such R & D and the need for its proper resourcing;
- stimulates debate on the future direction of the profession and the role of R & D within it.

RICS involvement

The RICS, as an organization, is committed to an R & D programme that combines long-term strategic studies aimed at determining the future shape and direction of the profession in the construction and property industries with targeting of those projects that support Institution policy developments. Towards the end of the 1980s the RICS encouraged, and in some cases sponsored, research in a wide variety of quantity surveying topics, including development appraisal, value management, estimating and bidding methods, risk analysis, life-cycle costing, expert systems, CAD, integrated databases and procurement systems.

In 1991, the RICS undertook a survey to help to determine the extent of these activities in the universities and published a report, *The Research and Development*

Strengths of the Chartered Surveying Profession: The Academic Base. The information included in the RICS report provided an overall profile of research activities, general areas of capability and specific research expertise, research links with the profession and details of external research contracts. Research and development may be classified into the following areas:

- *basic research*: experimental or theoretical work undertaken primarily to acquire new knowledge of the underlying foundation of phenomena and observable facts; undertaken without any particular application in mind;
- *strategic research*: applied research in a subject area that has not yet advanced to the stage where eventual applications can be clearly specified;
- *applied research*: the acquisition of new knowledge that is primarily directed towards specific practical aims or objectives.
- *scholarship*: work intended to expound the boundaries of knowledge within and across disciplines by in-depth analysis, synthesis and interpretation of ideas and information and by making use of rigorous and documented methodology;
- *creative work*: the invention and generation of ideas, images and artefacts including design; usually applied to the pursuit of knowledge in the arts;
- *consultancy*: the development of existing knowledge for the resolution of specific problems presented by clients, often within an industrial or commercial context.

R & D in the construction and property industries

R & D takes place in all industries, including the construction and property industries and their associated professions, and is important for the following reasons:

- Technical change is accelerating, and progressive businesses tend to quickly adopt new techniques and applications.
- R & D is inseparable from the well-being and prosperity of a country; as it is from the separate businesses within the country.
- Research and innovation are inseparable.
- The value of R & D to the construction industry cannot be overestimated. R & D is necessary to maintain international competitiveness and success, particularly as the craft-based traditions of construction diminish and the technological base expands.
- This background of constant change and challenges demands an effective R & D base to introduce change effectively and efficiently.

In respect of expenditure on R & D the UK construction industry lags far behind both its competitors overseas and other industries in the UK. Expenditure in the industry increased during the last decade, but in the UK still only amounts to only 0.65% of construction output. Construction companies, for example, contribute about 10% towards this sum, which is approximately one third of that of our competitors in France, Germany and Japan.

University R & D

Approximately 50 universities in the UK (and others abroad) offer courses in surveying. About half of these support quantity surveying programmes. Some of the quantity surveying courses are now taught in conjunction with other surveying specialisms or the different construction disciplines. An important part of any lecturer's activities is undertaking some form of R & D to support the students on undergraduate and postgraduate programmes. While some of this work is of a fundamental nature, seeking a better understanding and rationale for the principles that are often taken for granted, other work is done in collaboration with practice and industry to seek solutions, to improve or extend performance, or to address specific issues that have been identified.

For example, projects have been undertaken to benefit the profession from the development of the single European context, to advance the work of the surveyor in new areas of possible application, to harness new technology and to allow professional judgements to be applied with greater certainty. The RICS report referred to earlier indicated that funding for R & D in surveying generally amongst the different institutions amounted to over £15m. A large proportion of this included R & D being undertaken by quantity surveyors. The report stated that for research and development activities to fulfil their potential within an area of activity, it was necessary to stimulate a 'virtuous circle' of research. This is typified by the following features:

- There is sufficient awareness and confidence in the capabilities of the research base to encourage the profession to be prepared to commit funds to research.
- The flow of funds into research is perceived by the profession as being of great benefit to them.
- The profession sees the value in supporting research activities and is willing to continue to invest.
- There is a sufficient flow of funds into research in order to support and maintain a high-quality research base.

The above is the ideal. If a profession has achieved this 'virtuous circle' then it is very likely that it will be extremely well supported by research and development. It will be proactive, dynamic and forward-looking. In order to bring this about, one of the most important issues is how to overcome the barrier of the lack of knowledge of the R & D capabilities that already exist. For those who still doubt the wider benefits to be achieved from R & D, it is worth noting what the Centre Scientific et Technique du Bâtiment said in 1990: 'The stronger the research and development effort of a sector, the better its image even in a fragmented sector. Just look at the image of doctors!'

Changing role of the quantity surveyor

As with all professions, quantity surveying has evolved and will continue to do so for the foreseeable future. This evolution has been a response to changing demands and services expected from clients and the developing skills and

1950s	Quantity surveying practice associated with single-stage selective tendering, approximate estimating and final accounts.
1960s	Introduction of cost planning. Cut and shuffle. Standard phraseology. First use of computers for bill production.
1970s	Development of undergraduate courses in quantity surveying. Data coordination. Costs in use.
1980s	Emphasis on whole-life costing of projects. Coordinated project information. Value engineering. Alternative procurement systems. Project management.
1990s	Risk analysis. Wider application of computers. Decade of quality. Importance of information technology. Wider role of activities.

Fig. 15.1   Trends in quantity surveying practice.

knowledge base of practitioners, coupled with the far-reaching implications of information technology. Fig: 15.1 indicates some of the changes to the profession that have occurred since the middle of this century. Most of these changes have happened as a result of the pragmatic needs of practice in response to changes in the needs of clients and technology, rather than through any formal development or research.

As Fig. 15.1 shows, the role and work of the quantity surveyor has changed considerably, particularly over the past two decades. These changes in direction and practice are expected to be overshadowed by the accelerated developments that are likely to take place in the immediate future. A major theme of the report *QS 2000* is the changes facing the profession (see Chapter 2). The following represent some of the issues that, in the absence of appropriate and relevant R & D, may allow opportunities to be missed or to be ineffectively undertaken:

- blurring of professional disciplines, both within the surveying profession generally but also with other professional groupings;
- wider range of services offered to existing clients;
- application of surveying expertise to new markets;
- more extensive and intensive use of information technology to improve efficiency and effectiveness;
- changes in the professional structure;
- multi-discipline working and development;
- increased emphasis on continuing professional development;
- geographical dispersion of work to allow for the most economical methods of working;
- forecasted shift between professional and technician activities.

R & D in quantity surveying practice

Several of the larger quantity surveying practices have now established R & D sections as an integral part of their practices. This has been done in an attempt both to diversify and also to be at the leading edge of the profession. Some practices have been able to recoup fee income from work that can broadly be described as R & D. Others have joined in collaborative ventures with universities, become members of research advisory teams, or have allowed researchers access to non-sensitive data and information. Such activities are also able to provide a useful spin-off for public relations and publicity. R & D is therefore seen as being important for the following reasons:

- improving the quality of the service provided to clients;
- improving the efficiency of work practice;
- extending the services that can be provided;
- developing a greater awareness of new technologies;
- providing a fee-earning capability from R & D contracts;
- enhancing public relations and practice promotion.

Data collection and analysis

In some practices, R & D departments evolved from the routine collection of cost and contract information for regular use in cost planning and cost forecasting associated with new commissions. It is well understood by all surveyors that the best information available is that which is collected by the practice itself, relying only upon other sources, such as the BCIS or price books, as secondary sources of data.

Market trends

As construction costs and prices are associated with productivity and market conditions, it is important for surveyors to be informed of current trends in practice in order that clients can be properly advised. While the national published data is again valuable, this will form only a secondary source to the specific market information and trends that are retained by the surveying practice.

Practice expertise

A practice that has an extensive involvement with a particular type of construction project is able to provide detailed analyses on the different project components, so that the full extent of any changes in design or specification can easily be assessed. Such a practice is then able to develop a detailed expertise with a particular type of construction or procurement arrangement.

Objective and speculative R & D

It is also important to develop a research database prompted by the needs of the individual surveyors in order to allow for possible future surveying services to be developed. On this basis clients can be provided with information that in the future might generate new commissions for the surveyor. Development of this type may be undertaken with specific objectives in mind, or may be speculative, attempting to forecast changes in the market or in the demand for surveying services. It may also be undertaken to meet identified client needs for specific surveying services.

Fee-earning capability

In addition to the more usual surveying services, R & D activities are able to generate their own contracts for commissioned research. Some practices, for example, have developed a part of their own R & D expertise to publish the systematic collection and analysis of their data. Surveying practices are also able to submit R & D ideas or bids to government or research bodies for R & D contracts. This has in the past been done in association with a university or college as a collaborative venture.

The future

The future for the quantity surveying profession is intrinsically linked with a past analysis of its activities: the principle of reaping tomorrow what we have sowed today, together with a response to the changing forces in industrial and commercial society. The importance attached to education and training and continuing professional development is now well recognized if a profession is to be vibrant and dynamic. The importance of R & D within this forms an integral part in acquiring capability for the future. An effective R & D programme can only strengthen and assure the quantity surveyor's long-term future.

Bibliography

Brandon, P.S. (ed.) *Building Cost Modelling and Computers*. Spon, 1987.

Brandon, P.S. (ed.) *Quantity Surveying Techniques: New Directions*. Blackwell Scientific Publications, 1992.

Brandon, P.S. and McDonagh, N. 'Finding a framework for the future'. *Chartered Surveyor*, April 1991.

Centre for Strategic Studies in Construction *UK Construction Prospects 2001*. University of Reading, 1990.

Consensus Research *The Promotion of the Chartered Quantity Surveyor*. Consensus research, 1987.

Davis, Langdon and Everest *QS 2000: The Future of the Chartered Quantity Surveyor*. The Royal Institution of Chartered Surveyors, 1991.

Graves, R. 'Skills gap lets in predators'. *Chartered Quantity Surveyor*, June 1991.

IPRA Ltd *Future Skills Needs of the Construction Industries*. Report prepared for the Department of Employment, 1991.

Male, S. 'Professional authority, power and emerging forms of "profession" in quantity surveying'. *Construction Management and Economics*, 1990.

McDonagh, N. 'Future shock or future secured?' *Chartered Quantity Surveyor*, June 1991.

Nisbet, J. *Called to Account: Quantity Surveying 1936–86*. Stoke Publications, 1989.

Nisbet, J. 'Identifying the knowledge base'. *Chartered Quantity Surveyor*, June 1991.

Policy Studies Institute *Britain 2010: The PSI Report*. Policy Studies Institute, 1991.

RICS *RICS Fit for the 21st Century*. Junior Organization, The Royal Institution of Chartered Surveyors, 1991.

RICS *The Research and Development Strengths of the Chartered Surveying Profession: The Academic Base*. The Royal Institution of Chartered Surveyors, 1991.

The Core Skills and Knowledge Base of the Quantity Surveyor. RICS Research Paper No. 19, The
    Royal Institution of Chartered Surveyors, 1992.
Thompson, F.M.L. *Chartered Surveyors: The Growth of a Profession*. Routledge & Kegan Paul,
    1968.

Index